Science of El

Volume 7

Nuclear Power Technologies
Explained Simply

by Mark Fennell

© 2013

This book is part of the
Energy Technologies Explained Simply™ Series

Other Books in the Energy Technology Series

About the Book
Nuclear Power Technologies Explained Simply

Introduction

Nuclear Power Technologies Explained Simply is your one-stop resource for understanding everything related to Nuclear Power. This book is designed for citizens and policy-makers who want to become more fully informed regarding the science and technology of nuclear power.

All aspects of nuclear technology are explained simply enough for any reader to understand. At the same time, enough detail and data is provided for the reader to make intelligent decisions.

Answering All Your Questions

Within this book you will find answers to all of your questions related to nuclear power, including:

- How do nuclear power plants work?

- What are the main components and design options of nuclear power plants?

- What exactly happened at Three Mile Island and Chernobyl?

- How do we make nuclear power plants safer?

- What is radioactivity? How dangerous is each type?

- How do we shield ourselves from radioactive decay?

- How do we interpret the units of radioactivity? When I see a value related to radioactive decay, should I be concerned?

- How do we store nuclear waste safely for thousands of years?

- And many other questions related to nuclear power.

Extensive Data Tables

In addition, this book provides extensive data tables related to nuclear power, including:

- Complete list of radioactive isotopes. This table includes the decay type, the new atom created, and the half-life.

- Complete list of half-lives for all radioactive isotopes, listed in order of decay time.

- List of radioactive isotopes which transition through multiple decay sequences.

- Decay sequences for multiple decay isotopes.

- Summaries of radioactive decay.

- Melting points of nuclear fuel and fuel rods.

- Types of radioactive decay, including net effect on the atom, and shielding requirements.

- Dosage of absorbed radioactive decay and the resulting effect on human health.

- Suggested Nuclear Standards from the American Nuclear Society (ANS) and the Nuclear Regulatory Commission (NRC)

Chapter Topics

Nuclear Power Technologies Explained Simply consists of the following chapters:

7.1 Overview of Nuclear Power

The first chapter provides an overview of nuclear power. In this chapter you will get a broad overview of the process of creating energy from nuclear material. (Details will be discussed in subsequent chapters).

In this chapter you will also learn about the basic types of nuclear reactors and how they work.

The end of the chapter provides a high level overview of by-products and radioactivity.

7.2 Creation of Energy: Nuclear Fuel, Fission, Chain Reactions

Chapter two explains in detail how nuclear fuel is converted into energy. Fuels discussed include Uranium and Plutonium, as well as the composition of fuel pellets and fuel rods.

The actual processes of unleashing energy from the nucleus is done through fission, chain reaction, and critical mass. Each of these topics is discussed in sufficient technological detail to understand the processes.

The final topic in the chapter is about controlling the neutrons, which is the essential tasks in order to prevent meltdowns and explosions.

7.3 Operation of Nuclear Power Plants

Chapter three discusses the operation of nuclear power plants. In this chapter you will learn about all types of nuclear reactors. You will also learn the main components of any nuclear reactor. Finally, you will learn the main design options available for nuclear reactors.

7.4 Science of Meltdowns and Explosions

Meltdowns and explosions are the primary concerns for most citizens regarding nuclear power. Therefore chapter four explains the science of meltdowns and explosions in great detail.

The first section explains how meltdowns occur, supplemented with data on the melting points of nuclear fuel and fuel rods. The remaining sections explain the process of nuclear explosions. There are two types of nuclear explosions, and both are discussed in detail.

7.5 Three Mile Island

Chapter five is devoted to the case of Three Mile Island. In this chapter you will learn exactly what happened during this event. The series of events are described in a series of steps which are easy to follow. This description is supplemented by analysis of the incident.

7.6 Chernobyl

Chapter six is devoted to the case of Chernobyl. As with the chapter on Three Mile Island, in this chapter you will learn exactly what happened during the nuclear incident at Chernobyl. Through an easy to follow series of steps you will understand the events as they happened. This is supplemented by analysis of the incident.

7.7 Nuclear Accident in Fukushima Japan

Chapter seven is devoted to the recent nuclear incident in Fukushima, Japan. As with the previous chapters, you will learn what happened through an easy to follow step by step description of events.

7.8 Making Nuclear Power Plants Safer

In chapter eight you will learn all of the most important techniques for making nuclear power plants safer.

Chapter four explained the science of meltdowns and explosions. Chapter five through seven discussed the events (and the science) of each of the most famous nuclear power incidents. Now in chapter eight we will discuss all of the best ways to make nuclear power plants safer in the future.

There are several categories of nuclear safety. These safety categories include design, construction, control room, communications, maintenance, and training. This chapter discusses specific methods for making nuclear power plants safer related to each of those categories.

Note that the majority of these techniques are provided by the American Nuclear Society, Nuclear Regulatory Commission, and professors who specialize in nuclear safety.

7.9 By-Products and Radioactivity

Chapter nine explains the science of radioactivity. This chapter provides the basic science which will be applied to the remaining chapters of the book.

This chapter discusses alpha, beta, and gamma radiation. You will learn the nature and process of each type of decay, with examples. You will also learn about the many possible by-products of fission.

The last section is devoted to half-lives, where you will understand the basic concepts of half-lives, as well as the practical implications of various half-life values.

7.10 Health Issues of Radioactive Decay

Beyond meltdowns and explosions, the greatest concern of nuclear power is the effect radioactive decay has on human health. Therefore chapter ten discusses the health issues of radioactive decay in great detail.

Every aspect of radioactive decay on human health is explored. Topics include:

    a) routes of entry of radioactive decay into the body

    b) the penetration of each type of radioactive decay to enter the body

    c) mechanisms of each type of decay on the cells within the body

    d) overall health dangers of each type of radioactive decay.

The energy of the radioactive isotope is a major factor in how much damage will result. Therefore in this chapter we also discuss the energies of each type of radioactive decay. This includes the initial energies and the energies after impact.

Also, because gamma decay tends to be the most harmful, we discuss the energies of gamma decay. In particular we will discuss a new twist on the study of gamma decay: the energy of gamma as related to distance traveled.

7.11 Measuring Radiation

Chapter eleven discusses everything the reader needs to know regarding radiation measurements. This includes how radiation is measured, units of measurements, and biological effects.

There are many types of radiation measurements. These can be broadly grouped into physical properties, biological effects, and dose equivalents. The first three sections of this chapter explain each of these types of radiation measurements so that you can understand them clearly

The concept of dose equivalents is subtle, and we must use our logic when interpreting dose equivalent data. Therefore we will discuss dose equivalents in great detail.

Within each of these sections I provide a complete list of units used for radiation measurements. These are defined and explained, with additional notes that may help the reader.

Many people are concerned about the biological effects of radiation. Therefore I have a section which gives the reader a quick guide to dosage of radioactivity and the resulting biological effects.

The final sections of this chapter discuss devices which measure qualitative amounts of radiation (versus absolute amounts), such as the Geiger Counter, the Scintillator, and the Film Badge. We will discuss how each works, where they are most effectively used, and their limitations.

7.12 <u>Storing Nuclear Waste</u>

The final chapter of this book corresponds to the final activity of nuclear power: how to store radioactive nuclear waste for long periods of time.

We begin this chapter with a discussion on the steps required to store nuclear fuel, from depleted fuel rods in the hot reactor core to the final storage underground.

We then look at the possible dangers to the stored nuclear waste. We first examine the possible dangers from the nuclear material itself, within the container, and how to minimize those dangers. Then we look at the outside dangers to the container, such as from tornadoes, water, and terrorism. We will also discuss how to prevent those dangers from occurring.

Then we will discuss the repository at Yucca Mountain. We will discuss in great detail why the Yucca Mountain site has been chosen for the federal repository of nuclear waste, focusing particularly on the advantages of the location and the design of the facility.

We will conclude this chapter with a brief look at the future of nuclear waste technology.

<u>Comprehensive Data in the Appendix</u>

At the end of the book you will find a comprehensive set of data. This is the most comprehensive and most complete set of data related to nuclear power you will find anywhere.

The first table is a complete list of radioactive isotopes. In addition to the isotope, this table also provides the decay type, the new atom created, and the half-life. The first lines are as follows:

Element	Isotope	Decay Type	Becomes	Half-Life
1 (Hydrogen, H)	H–3	beta	He(2)–3	12 years
4 (Beryllium, Be)	Be–10	beta	B(5)–10	1,600,000 years

This table was created by using multiple sources, because no single source provided all the data. Now, in this book, you have the most complete set of data on radioactive decay anywhere.

The second table is a complete list of half-lives for all radioactive isotopes, listed in order of decay time. A small sampling of the data items for half-life are as follows:

Isotope	Decay Type	Half-Life
Polonium-214	alpha	.00016 seconds
Potassium-43	beta, gamma	22.3 hours
Cesium-134	beta, gamma	2.1 years
Plutonium-239	alpha, gamma	24,110 years
Neptunium-237	alpha, gamma	2,140,000 years
Uranium-238 (99.3%)	alpha	4,500,000,000 years

Note that some radioactive isotopes require multiple steps in order to become stable. Therefore I provide a complete list of all radioactive isotopes which require multiple decay steps. A small sampling of the data includes the following:

Original Isotope	# steps	Total Decay
Lead-210 (Pb-210)	3	2 beta; 2 gamma; 1 alpha
Polonium-218 (Po-218)	7	3 alpha; 4 beta; 4 gamma
Thorium-234 (Th-234)	13	7 alpha; 6 beta; 8 gamma

I also provide a unique set of tables based on multiple decay sequences. After reviewing the decay sequences, I discovered that I could place ALL multiple decay sequences in just four simple tables! This greatly simplifies the organization and understanding of multiple decay sequences. Of course, you will not find this set of tables in any other resource!

Other data in the Appendix include summaries for radioactive decay, and lists of suggested reading for nuclear safety standards.

Ultimate Resource on Nuclear Power Technology

In total, this book is the ultimate resource for citizens and decision makers on nuclear power technology. *Nuclear Power Technologies Explained Simply* will answer all your questions related to nuclear power.

This book will guide you through all the science and technology. Every aspect of nuclear power is explained simply enough so that anyone can understand regardless of background, yet enough technical detail is provided for the reader to make accurate and informed decisions.

About the Science of Electricity Series

<u>Purpose of this series</u>

The books in the *Science of Electricity* series are designed to be a resource for citizens, students, and legislators to learn about all aspects regarding the science of electrical power.

The ultimate goal of the series is to enable the people to make informed decisions on practical energy questions. The secondary goal is to serve as introductory guides for students embarking on careers with energy technologies.

The books in the series, taken together, are designed to answer all of your practical questions on each type of energy technology. The books answer common questions such as:

- How can we increase the efficiency of solar cells?
- How do I select the size my solar array?
- What do I need to know when installing a wind turbine?
- How effective are the clean coal technologies?
- How can we prevent grid failures?
- Do power lines cause cancer?
- and many other energy technology related questions…

<u>Specific Goals</u>

There are numerous technologies described in these books. Yet for each technology I sought out the answers to the following questions:

1. How does the technology work?
2. What are the most important terms and abbreviations?
3. What are the advantages and disadvantages?
4. What is the environmental impact? How can it be improved?
5. What are the safety hazards, and how can they be reduced?
6. What are the most important practical tips?
7. What facts comprise the most important data?

<u>Technical Discussions Explained Simply</u>

The books in the series must necessarily be technical to some degree. Electricity is a practical technology, and therefore we must understand the technical aspects if we want to make wise decisions. Yet the discussions are always aimed at the citizen or policy maker.

The books in this series explain the principles of electricity as simply as possible, using ordinary English (no engineering jargon), and highlighting the most important points of each technology. Main concepts and facts are emphasized with the use of lists, tables, diagrams, and summaries.

I do not expect any reader to have a background in science, yet I offer enough facts and details so that you can have an accurate understanding of all related technologies. I provide enough technical details and enough data for the reader to make informed decisions.

Objectivity

I have tried my best to be as objective as possible. Whereas many other authors of energy books have an agenda, I have no desire to promote one industry over another. I have no desire to promote one technical solution over another. In this endeavor, I have tried to be an objective scientist.

Accuracy of Data and Summaries

I never relied solely on the conclusions of other researchers. Instead, I performed many other tasks to ensure that all conclusions were accurate. I examined primary data whenever possible. I have read the fine print on how research was obtained. I have also checked the accuracy of the conclusions by finding at least three distinct sources for each fact.

It is only after such rigorous investigations that I created data tables and wrote summaries for these books. Because of the labor I performed in fact checking, I fully stand by every statement and every data item written in this series of books.

Conclusion

For all the reasons above, I offer this series of books. My goal is to inform you on the basic concepts of all the technology and all of the issues related to electricity so that you can make realistic decisions.

Remember that there are no perfect solutions, there are only choices. I hope that this series of books will assist you in making those choices for your community.

Mark Fennell

Table of Contents

Table of Contents: Detailed

7.1
Nuclear Power Basics

Introduction

Energy from Nuclear Power

We must be clear on the basics of nuclear power and not lose perspective: nuclear power is a high tech way of boiling water.

Nuclear power is based on the energy stored in the nucleus of an atom. Splitting the nucleus of an atom apart will unleash great energy. This energy is used to heat water, thereby turning the water into steam. This steam flows to the turbines. The steam pushes the blades of the turbines, which operates the electrical generators, and the generators create the electricity.

Raw Material/Nuclear Fuel

The atoms used for nuclear power are chosen because they are fissionable (able to split easily). There are two basic fuels used for nuclear reactors: Uranium-235 and Plutonium.

Nuclear fuel must be prepared more than any other fuel. This can be a challenging step because Uranium-235 is not common. Uranium-235 is only .7% of all the forms of Uranium. Therefore U-235 must be refined from all the other forms of Uranium before we can use it as fuel.

Regarding Plutonium, this element does not exist naturally. Therefore we must create Plutonium before we can use it as fuel.

Creation of Energy: Fission and Neutrons

Fission is the process of splitting the nucleus of an atom. In order to split an atom, we fire a neutron at the nucleus. The nucleus splits, and energy is released. If a type of atom splits easily, we call this atom "fissionable." Whether a material is fissionable or not depends on the particular isotope of the element.

During a fission process we get three important items: energy, extra neutrons, and the creation of smaller atoms. Energy is primarily what we are after. The extra neutrons help us create further fission reactions. The smaller atoms we create are considered the by-products of the fission process. These by-products are sometimes radioactive and may cause health problems.

Basic Nuclear Power Plant: The "BWR"

The basic nuclear power plant is called the Boiling Water Reactor, abbreviated BWR (Figure 7.1).

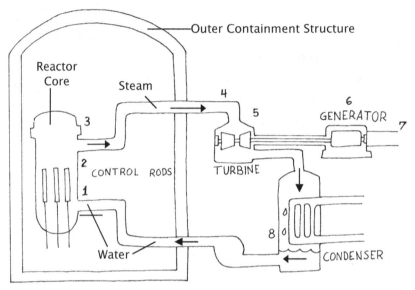

Fig. 7.1 Boiling Water Reactor (BWR)

In brief, the process of the Boiling Water Reactor is as follows:
1. Fission process produces energy.
2. This energy is put into water.
3. Water becomes steam. (Energy from the nuclear fission heats the water, making it into steam)
4. Steam flows to the turbine, causing turbine blades to rotate.
5. Turbine rotating works the electric generator.
6. The generator creates electricity.
7. That electricity gets connected to our homes.
8. To continue the process, the steam is cooled, condensed into liquid water and circulated back. This same water is then run by the next nuclear reaction, absorbs more heat from the nuclear energy, and the process continues.

Boiling Water Reactor with Heat Exchange

Overview

The boiling water reactor with heat exchange is a very common variation of the basic nuclear power plant process described above. This process uses two liquids, usually two sources of water, rather than just one source of water as in the reactor above. The first liquid gives its heat to the second liquid, in a process called "heat exchange."

It is important to understand the distinction between the two waters, and how the heat is exchanged. Water #1 is self-contained. It never leaves the power plant system. Water #2 takes the heat from water #1. Water #2 becomes the steam that actually pushes the turbine.

Note that the two waters do not physically interact. One tube is placed inside another tube, Heat is transferred because the waters are close, but the waters do not mix. No nuclear radiation enters water #2.

BWR with Heat Exchange: Details of Process

The Boiling Water Reactor with Heat Exchange creates power in the following steps: (Figure 7.2)

1. Fission process splits atoms and produces energy.

2. This energy is put into water. (This is Water #1. It is contained water.)

3. Water becomes steam. (Energy from Nuclear Fission heats the water, making it into steam.)

4. Water #2 comes from a river or from a secondary stored water supply.

5. Steam from water #1 heats water #2, which is in a different container. The waters do not touch, but heat is passed from water #1 to water #2.

6. Water #2 (now as steam) flows to the turbine, causing the turbine blades to rotate.

7. Turbine rotates which operates the electric generator.

8. The generator creates electricity.

9. Water #2, after pushing the turbine blades, then goes to an area to cool before being put back into the secondary water supply.

10. Water #1 is also reused, sent through the reactor core to absorb more nuclear energy from the fission reactions, and continues the process.

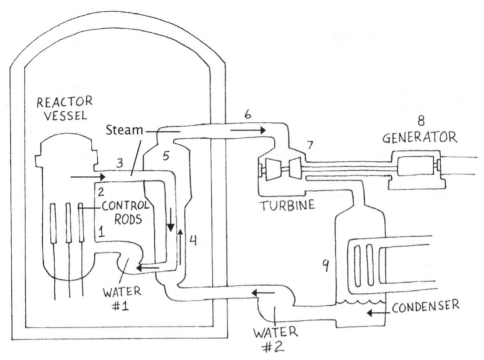

Fig. 7.2 Boiling Water Reactor with Heat Exchange

Power and Energy from a Nuclear Power Plant

One of the greatest advantages of nuclear power is the amount of power that we can get from just a small amount of fuel. A single fuel pellet of Uranium is less than one cubic inch. Yet just one of these tiny uranium pellets can give us as much energy as: one ton of coal, two tons of wood, three barrels of oil, or 17,000 cubic feet of natural gas.

The energy that we get from a fission reaction is from the binding energy of the nucleus. When we split that nucleus, the binding energy is released. Much of the energy is released in the form of kinetic energy of neutrons and atoms. Each smaller atom and each neutron resulting from the fission reaction travels very fast. The initial velocity of these particles is primarily due to the binding energy released from the fission reaction. Controlling the velocity of these particles will control the amount of power produced. The exact amount of power from a nuclear power plant depends on three factors:

1. The amount of fissionable material in a small area.

2. How many neutrons we allow free versus how many neutrons that we prevent from hitting other atoms.

3. The kinetic energy of the fission by-products.

We will discuss each of these factors in detail in later chapters.

Overview of By-Products and Radiation

Introduction

Society would not be talking about by-products of nuclear power except for this two-part reason: many of the by-products are radioactive, and radioactivity affects the health of humans.

Note that radioactivity is natural. There are over a hundred naturally occurring radioactive isotopes. They exist with us or without us. If we had never built any nuclear power plants, there would still be some radiation from natural radioactivity.

It is important to realize that after a nuclear plant has been running for a while, there will be many by-products. Some of these by-products will be radioactive, some will be stable. Furthermore, of the radioactive by-products, some will become safe within a few minutes, while others will require years of containment.

Radioactivity: Transforming One Element into Another

The entire reason that radiation exists is that some elements want to transform into another element. This is the entire purpose of an atom wanting to do the radiation process. An atom which was by nature put together in an unstable way will want to become stable. This means that the nucleus will change. When the nucleus changes, the identity of the element changes.

The process of one atom becoming another atom is called radioactive decay, and any element which wants to turn into another element is called a radioactive isotope. Each of these terms is used to describe the process of an unstable atom turning into a more stable atom. The basic process for transforming one element to another is the same for all types of radiation:

1. Something is jettisoned from the nucleus. (This is the "decay" of radioactive decay.)

2. The result is a change in the nucleus.

3. A change in the nucleus means a change in the very identity of the element.

Identity of the Atom: The Element and Isotope

The identity of an atom is specified by the element name (which is determined by the number of protons), and then by the isotope number.

The number of protons defines an element. For example, Uranium is Uranium (not any other element) because it has 92 protons. Going further across the periodic table, an atom with 93 protons, by definition, is Neptunium, and an atom with 94 protons is Plutonium.

An isotope is the same type of atom, but differs only in the number of neutrons. Furthermore, isotopes are described by their numbers. For example, Uranium-235 is one type of isotope of Uranium. Other isotopes include U-238 and U-234. The isotope number, such as 235, tells us total number of protons and neutrons in the nucleus of that atom.

Types of Radioactive Decay

There are three types of radioactive decay:

Name	What it is
Alpha particle	Helium atom
Beta particle	Electron
Gamma ray	electromagnetic wave

The Net Effect of Alpha decay is:

1. Helium atom is jettisoned from the nucleus.
2. Atomic # decreases by two. Thus the identity of the element changes to the *left* by *two* elements on the periodic table.
3. The isotope # decreases by four.

The Net Effect of Beta Decay is:

1. An electron is jettisoned from the nucleus.
2. Atomic # increases by one. Thus the identity of the element changes to the *right* by *one* element on the periodic table.
3. Isotope # remains the same.

The Net Effect of Gamma Decay is:

1. A burst of a short electromagnetic wave is given off.
2. Gamma affects the atom by changing its energy.
3. Gamma radiation does *not* change the identity of the atom.

Summary

1. Nuclear power is a high-tech way of boiling water.

2. Nuclear power is based on the energy stored in the nucleus of an atom. The energy that we get from a fission reaction is from the binding energy of the nucleus. When we split that nucleus, the binding energy is released. The energy released is used to boil water. The resulting steam is used in a steam turbine and generator to create electricity.

3. The fuel for nuclear power is usually Uranium-235 or Plutonium.

4. During a fission process, we get three important items:
    a. energy (primarily what we are after)
    b. extra neutrons (helps us create further fission reactions)
    c. the creation of smaller atoms (which are sometimes radioactive)

5. The amount of power from a nuclear power plant depends on:
    a. The amount of fissionable material in a small area.
    b. How many neutrons we allow free
    c. Kinetic energy of the fission by-products.

6. A single fuel pellet of Uranium is very small, less than one cubic inch, yet just one of these tiny Uranium pellets can give us as much energy as one ton of coal or three barrels of oil.

7. The basic nuclear power plant is called the Boiling Water Reactor, abbreviated BWR. In a BWR, the nucleus of an atom is split apart, which unleashes a great amount of energy. This energy is used to heat water, thereby turning the water into steam. This steam flows to the turbines, which operates the electrical generators, and creates electricity.

8. The boiling water reactor with heat exchange is a very common variation of the basic nuclear power plant process described above. This process uses two liquids, usually two sources of water, rather than just one. The first liquid gives its heat to the second liquid, in a process called heat exchange. The water which flows past the radiation is always contained, never leaving the reactor.

9. Radioactivity is the process of an unstable atom transforming into a stable atom. Radioactivity is a natural process.

10. The basic process of radioactivity is:
     a. Something is jettisoned from the nucleus. (This is the "decay" of radioactive decay.)
     b. The result is a change in the nucleus.
     c. A change in the nucleus results in a new identity of the element.

11. The identity of an atom is specified by the element and then the isotope number. The element is defined by the number of protons. The isotope number is the total number of protons and neutrons.

12. The three types of radioactive decay are alpha, beta, and gamma.

13. The net effect of Alpha decay is: A helium atom is jettisoned from the nucleus, the identity of the element changes to the left by two elements on the periodic table, and the isotope # decreases by four.

14. The net effect of Beta Decay is: an electron is jettisoned from the nucleus. The identity of the element changes to the right by one element on the periodic table. The isotope # remains the same.

15. The net effect of Gamma Decay is: A burst of a short electromagnetic wave is given off. Gamma affects the atom by changing its energy. Gamma radiation does not change the identity of the atom.

7.2
Creation of Energy:
Nuclear Fuel, Fission, & Chain Reactions

Introduction

Nuclear power is based on the energy stored in the nucleus of an atom. When we split that nucleus, the binding energy is released, which is then used to boil water. The resulting steam is used in a steam turbine and generator to create electricity.

For nuclear power we need to process our raw materials before we can use them. There are two basic nuclear fuels that we use for our fission reactions: Uranium-235 and Plutonium. Both of these fuels require processing before we can use them. The raw material for both is a supply of Uranium.

Uranium

A natural supply of Uranium is made of several isotopes, notably Uranium-238 and Uruanium-235. U-235 is fissionable and can be used as fuel for nuclear power, but U-238 is not fissionable. Therefore, we must separate the U-235 from the rest of the Uranium.

However, there is far less U-235 than U-238 in any ore of Uranium which makes separation a difficult process. In nature any sample of Uranium is approximately: U-238 = 99.2%, U-235 = .7%, U-234 = .006%. In general, it is difficult to separate isotopes of the same element. It is also difficult to separate isotopes when there is such a small percentage to begin with. Therefore, refining Uranium is a difficult and tedious process.

Refining Uranium usually involves specialized equipment which can separate U-235 from U-238. In addition, the separation process is usually repeated multiple times in order to get sufficient quantities of U-235.

When a sample of Uranium has been refined to a higher percentage of U-235 than the natural ore, then that sample is called "enriched Uranium." A nuclear power plant requires Uranium fuel that is enriched to 2%-5% U-235.

Note that this value is far below that of a nuclear weapon. A nuclear weapon requires fuel that is 90%–95% U-235. A Uranium sample of this percentage is called "weapons grade". This requires much more processing and is much more expensive.

Plutonium

It would be nice to use Uranium-238 since it is far more abundant than Uranium-235. However, U-238 is not fissionable. Instead we can convert Uranium-238 into Plutonium. Plutonium is fissionable, therefore we can use this Plutonium as fuel in our nuclear reactors.

Note that Plutonium does not occur naturally. Plutonium is an artificial element, and it is only created from Uranium-238. The process of converting Uranium-238 into Plutonium is done through two steps of beta decay. This means that we take two electrons away from the nucleus, thereby creating Plutonium.

We start the process by adding a neutron to Uranium. This makes the Uranium unstable, and therefore wanting to go through beta decay. Beta decay then occurs naturally, twice. (Recall that beta decay is losing an electron from the nucleus.) This process of beta decay, in two steps, creates our Plutonium.

In the first beta decay, Uranium (92 protons) becomes Neptunium (93 protons). In the second beta decay, Neptunium (93 protons) becomes Plutonium (94 protons). You might see the equations written as this three step process:

1. U-238 + n = U-239 (Start the process by adding a neutron)
2. U-239 = Np-239 + e- (Uranium becomes Neptunium, loses e-)
3. Np = Pu-239 + e- (Neptunium becomes Plutonium, loses e-)

The conversion times are relatively short. U-239 converts to Neptunium quickly, with a half-life of 23 minutes. Neptunium converts to Plutonium relatively quickly, with a half-life of two to three days.

Several isotopes of Plutonium can be created by this process. The most common Plutonium isotopes used as fuel for nuclear reactors are Pu-239, Pu-240, and Pu-241. From these isotopes we can use Plutonium in our nuclear reaction power generation.

Fuel Pellets and Fuel Rods

After the fuel is refined then it is put into pellet form. The volume of a fuel pellet is a cubic inch or less. These pellets are then placed into fuel rods. The rods are made of metal, usually Zirconium.

When Uranium is used as nuclear fuel it can be either be elemental Uranium (U), or Uranium Dioxide (UO_2). The amount of U-235 is very important. For an effective chain reaction to take place, the Uranium in each pellet must be at least 2% U-235. (The rest of the pellet is U-238, which is not fissionable.) The higher the percentage of U-235 in the pellet, the more easily fission will take place.

When Plutonium is used as the nuclear fuel it can be one or more of many Plutonium isotopes. Each isotope has its own characteristics of being useful as nuclear fuel. Nuclear engineers estimate the percentage of each Plutonium isotope for their particular needs.

Fission and Energy

Fission Basics (Figure 7.3)

A nucleus is made of protons and neutrons. The protons and neutrons are held so tightly that it is difficult to separate. However, if you manage to break the nucleus apart then the energy holding the nucleus together is unleashed.

From a fission process we get three important items: energy, extra neutrons, and the creation of smaller atoms. Energy is primarily what we are after. This is the entire purpose for us to do fission. The extra neutrons help us create further fission reactions. The smaller atoms that we create are considered the by-products of the fission process.

First we need to start the fission process. Just as we need a match to start a fire, we need something to start the fission process. We start the fission process by firing a neutron at an atom. When the neutron hits the atom, then the atom splits and we get our energy.

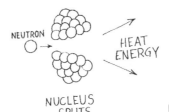

NEUTRON
HEAT ENERGY
NUCLEUS SPLITS

Figure 7.3 Fission Reaction

<u>Examples of Fission Process</u>

The following is a typical example of a fission process. We split a Uranium atom, creating Barium and Krypton. Energy is released, and three extra neutrons are ejected. In chemical equation form, this might be written as: $1\,n + U\text{-}235 = Ba\text{-}142 + Kr\text{-}91 + 3\,n + Energy$.

Note that we can have other smaller atoms than Barium and Krypton. After a fission has occurred we are not always sure of exactly what atoms we broke the Uranium into. As long as the number of protons in each smaller atom adds to 92 (the number for Uranium), then we can have any combination of smaller atoms. This is important to note because of the concern of radioactivity of the by-products. How can you measure the radioactivity of the by-products if you're not sure what by-products you have in the mix?

Chain Reaction and Critical Mass

<u>Chain Reactions and Neutrons (Figure 7.4)</u>

Theoretically, we could keep firing neutrons at each atom, and create energy from fissions one by one. However, this would be like holding a match to every molecule of wood we wish to burn. There is a better way. When a nucleus is split, a few neutrons are thrown out. These neutrons can then be used to split other atoms. The series of these fissions is called a "chain reaction."

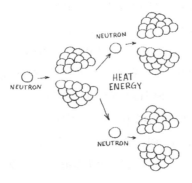

Figure 7.4 Chain Reaction

<u>Critical Mass</u>

However, there is one practical point to make this work. Neutrons are small, and a Uranium atom is mostly empty space. Therefore, it is difficult to hit the nucleus with the tiny neutron. In order to get enough direct hits, we must have enough fissionable material in a small space. Only when there is enough fissionable material in a small space for the free neutrons to easily find and hit those fissionable atoms will we be able to have a continuous fission process. The amount of material to be effective is called the "critical mass."

Controlling the Neutrons

<u>Introduction</u>

The problem with nuclear power is not how to get more energy out of the process, but rather how to keep too much energy from being created at one time. The way to control the reaction is to control the neutrons.

With each fission reaction more neutrons are produced. These neutrons split more atoms, which produce more neutrons, resulting in more energy and more neutrons.

In the following equations, watch the chain reaction grow:

Step 1: 1 n + 1 U–235 = 2 atoms + 2 n + 1 unit energy
Step 2: 2 n + 2 U–235 = 4 atoms + 4 n + 2 units energy
Step 3: 4 n + 4 U–235 = 8 atoms + 8 n + 4 units energy
Step 4: 8 n + 8 U–235 = 16 atoms + 16 n + 8 units energy
Step 5: 16 n + 16 U–235 = 32 atoms + 32 n + 16 units energy
           and so on....

To a certain extent, we do want this. This is the chain reaction which makes the production nuclear power effective. However, if we did not control the amount of neutrons in some way then the chain reaction would quickly grow to a greater extent than we could handle. If we didn't control the neutrons during a chain reaction then we could create an explosion.

<u>Control Rods</u>

A chain reaction process means the production of many more neutrons thrown off, many more fission reactions, and much more energy released. If we did not control the neutrons, then the chain reaction would produce more energy than we could handle. That is why we must have control rods. These control rods absorb some of the neutrons and thus keep the chain reaction to a level that we can work with. The control rods keep the energy at a level that is safe, not explosive.

This is important enough to repeat: the number of free neutrons is the primary factor in determining the amount of energy produced in a nuclear power plant. When we control the number of free neutrons, we can manage the amount of energy released. The control rods and safety rods are the devices which absorb neutrons and thus control the reactions.

Chapter Summary

1. There are two main fuels used in nuclear power: Uranium-235 and Plutonium.

2. The majority of Uranium is U-238. However, U-238 is not fissionable.

3. Uranium can be refined to get higher percentages of U-235, then used as fuel. This form of nuclear fuel is known as enriched uranium.

4. Plutonium is man-made. U-238 is the starting material. The process involves adding a neutron, followed by two beta decays.

5. Nuclear fuel is put into the form of pellets and held in place by a fuel rod.

6. The energy from nuclear power comes from splitting the atom, a process called fission.

7. We start the fission process by firing a neutron at an atom. When the neutron hits the atom, then the atom splits.

8. From a fission process, we get three important items: energy; extra neutrons; the creation of smaller atoms.

9. A chain reaction is a continuing process of multiple fission reactions. However, in order for this to work, we need enough fissionable material in a small space, an amount known as critical mass.

10. The problem with nuclear power is not how to get more energy out of the process, but rather how to keep too much energy from being created at one time. In essence, an uncontrolled chain reaction could create an explosion.

11. We control the amount of energy created in a nuclear reaction by controlling the amount of free neutrons and by controlling how fast the neutrons fly.

7.3
Operation of Nuclear Power Plants

Introduction
There are several types of nuclear power plants. There are also several design options for any nuclear power plant. We will discuss all the design possibilities in this chapter.

Main Types of Nuclear Reactors

Introduction
There are four main types of nuclear reactors:
1. Boiling Water Reactors (BWR)
2. Boiling Water Reactors with heat exchange (2 water sources)
3. Pressured Water Reactors (PWR)
4. Breeder Reactors, or Breeders

Boiling Water Reactors (BWR)
The BWR is the basic type of reactor (Figure 7.1). Energy is released from a fission reaction. This energy is used to boil water into steam. This steam flows to the turbine and thus operates the generator.

Boiling Water Reactors with Heat Exchange (2 water sources)
This is a variation of the BWR above, and is also very common (Figure 7.2). Energy is released from a fission reaction. This energy is used to boil water into steam. The heat from the steam is transferred to a second water source, in an area called the heat exchanger. The second water then flows to the turbine and operates the generator. The first water circulates back to the core to get more energy from fission reactions.

Pressured Water Reactor (PWR)
The pressurized water reactor (PWR) is very similar to a Boiling Water Reactor with heat exchange. However, in the pressurized water reactor, the water in loop 1 (the water flowing through the core) is pressurized. Due to the pressurizer on loop 1, the water remains as hot water, rather than steam. That is the only difference between the PWR and the BWR.

These reactors can then be smaller. Pressurized water reactors (PWR) are commonly used to power ships in the U.S. Navy (because these reactors are smaller). Some of the major power plants in the United States are also PWRs.

<u>Breeder Reactors, or Breeders</u>

The Breeder Reactors take a supply of Uranium-238 and turn it into Plutonium. In that context, we are "breeding" a fissionable material: Plutonium. Then the reactor goes through the same fission process as other nuclear reactors, this time using Plutonium atoms as our fissionable fuel. Breeder reactors are also often simply called "breeders."

Recall how plutonium is created: We start the process by adding a neutron to Uranium-238. This makes the Uranium unstable, and therefore wanting to go through beta decay. Beta decay then occurs naturally. This process of beta decay, in two steps, creates our Plutonium. From there we can use Plutonium in our nuclear reaction power generation.

Plutonium Isotopes used in Breeders

A few words should be written on the Plutonium isotopes used in Breeders. The most important of the Plutonium isotopes is Pu-239. This is the isotope which we create from Uranium-238. However, additional isotopes of Plutonium can be created by subsequent steps.

Recall that in order to create Pu-239, we start with U-238 and add a neutron. Similarly, if a neutron is added to Plutonium-239 then that atom will become Plutonium-240. This process can continue. With every added neutron, we get a new Plutonium isotope. In the end, it is possible to get not only Pu-239 from our Breeder Reactor, but we can also get several heavier isotopes of Plutonium.

The possible isotopes of Plutonium in this creation process can be: Pu-239, Pu-240, Pu-241, Pu-242, Pu-243, and/or Pu-244. Each Plutonium isotope has its own characteristics of being useful as nuclear fuel. Each has its own characteristics of radioactive decay. Nuclear engineers will often separate these isotopes of Plutonium in order to get the ideal percentages of each. The ideal percentages vary, depending on the goals of the engineers.

Nuclear Plant Equipment

Introduction

It is important to understand the main pieces of equipment within a power plant. In this section we explain each of the main terms regarding nuclear power plant operation.

Reactor

A reactor is a power plant which uses nuclear fission and steam to create electricity. The term "reactor" refers to the entire plant. Thus, the reactor can be any of those types of nuclear power plants listed above.

Reactor Core, or Core

The core is the area where the fuel is stored and where the fission reactions actually take place.

Fuel Rods

The fuel rods are the fissionable material. The fuel rods are the main part of the reactor core. The fuel is kept in solid form, usually in pellets, and these pellets are held by a container. The container holding the fuel pellets is the fuel rod.

Control Rods

Control rods are rods which absorb the neutrons, thus limiting the nuclear reactions. The control rods are usually made of cadmium steel, a material which absorbs the neutrons easily. The control rods are essential in maintaining safe operation in the core.

The control rods can be raised up and down. When the control rods are raised, they are out of range of the neutrons. When the control rods are down, they can absorb the neutrons. At the start of the fission process, the control rods are raised. After the process starts the control rods are lowered. Engineers constantly monitor the chain reaction, and lower or raise the control rods as needed to control the energy from the fission reactions.

Safety Rods

In addition to the control rods, most plants have safety rods. These are identical to control rods, made of the same material and used for the same purpose. However, these rods are only used in an emergency. They are kept raised all the time. Should the need arise, the safety rods are dropped. These safety rods can have a great and immediate effect in absorbing neutrons, and therefore can control the reaction in an emergency.

Moderator

Another way to control the chain reaction is to slow the neutrons. This is done with the moderator. The fuel rods sit inside this moderator at all times.

The moderator is usually water. The fuel rods (and the control rods when lowered) sit inside this body of water. The neutrons are slowed just a bit by the moderator before proceeding to their next fission. Therefore, this water "moderates" the reactions.

A moderator can also be graphite. In this case, the moderator is a series of graphite rods. In this situation, all rods (meaning all fuel rods, all control rods, and all moderator rods) are placed close together.

Note also that the temperature of the moderator is important. The reactions will naturally heat up the water in the moderator. If the water in the moderator is not kept cool enough, then heat will build. This heat can make the fission reactions proceed that much faster. Therefore, moderator water must be kept cool enough so as to not contribute to a run-away chain reaction.

Heat Exchangers

In most nuclear power plants there are two loops of water. The first loop of water flows directly past the core and absorbs the heat directly from the fission. The second loop of water is what actually pushes the turbine blades. The energy from the first water is transferred to the second water in the heat exchanger.

In the heat exchanger one pipe is contained within the second pipe. The waters do not mix, but the heat is transferred because they are close.

Cooling Towers

Cooling towers are towers which cool the water. These are the huge concrete towers that we associate with nuclear power plants. In the cooling towers, the heat of the water is transferred to the heat of the air. The hot air leaves the cooling tower, and the cool water remains to be reused.

Note that most of the nuclear reactor is small in comparison to the cooling towers. The reactor core is usually just a modest building. The tall towers exist only to cool the steam into water to be reused.

Cooling System

The term "cooling system" is vague. A cooling system can refer to any system in the power plant which exists to cool some part of the power plant.

A nuclear power plant has several different cooling systems. Some of the cooling systems are: 1) the heat exchange of the two waters; 2) the cooling towers; 3) the moderator; and 4) the various cooling systems to cool equipment (as there would be for any power plant). In addition to those cooling systems, there are 5) all the emergency cooling systems. Therefore, the term "cooling system" is rather vague, and should be specified in discussions.

Coolant

A coolant is any liquid or gas which is used to take heat away from an area. In general, the coolant is allowed to flow past a hot substance (pipes, equipment, fuel). The heat is transferred to the coolant, and the coolant is removed from the area. The most common types of coolants are: water, air, liquid metal (usually liquid sodium), carbon dioxide, or helium.

When discussing coolants we should specify the coolant system, because different coolant systems often use different coolants. (See "coolant system" above.)

Design Options for Nuclear Reactors

In addition to the basic types of nuclear reactors, there are various choices of options. In many ways this is like shopping for a car: 2 doors or 4 doors? Blue or Red? Just as a car needs a door, we do need the items below. However, there are design choices.

Also note when you read of types of reactors, the authors of the report may use any of these terms to classify the type of reactor. Just as you could classify your car as "4 door" or as "red", writers may classify the nuclear reactor by any of these items below.

The primary design options are:

1. Fuel Type: enriched Uranium (2% to 5% U–235) or Plutonium

2. Moderator: water or graphite

3. Coolant: water, air, or liquid metal

4. Velocity of Neutrons: Thermal (1 mile/second)
   or Fast (10,000 miles/second)

5. Use: Power generation, breeder and power generation, research, or atomic bombs.

Chapter Summary

1. There are four main types of nuclear power plants:
   a. Boiling Water Reactors (BWR)
   b. Boiling Water Reactors with heat exchange
   c. Pressured Water Reactor (PWR)
   d. Breeders

2. A nuclear power plant has these main parts:

   a. Reactor: The entire power plant is a "reactor."

   b. Reactor Core: The "core" is the where the fuel is stored and where the fission reactions actually take place.

   c. Fuel rods: The fuel rods are the containers which hold the fissionable material. The fuel rods are the main part of the reactor core.

   d. Control Rods: Control rods are rods which absorb the neutrons, thus limiting the nuclear reactions.

   e. Moderator: The moderator is a liquid (usually water) or solid graphite which surrounds the reactor core. The moderator's purpose is to absorb heat and keep the reaction controlled.

   f. Safety Rods: Safety rods are identical to control rods except that these are used only in emergencies.

   g. Heat exchangers: In most nuclear power plants there are two loops of water. The first loop of water flows directly past the core and is self-contained. The second loop of water goes past the first one, absorbing the heat from the first loop. That process is the heat exchange.

   h. Cooling Towers: Towers which allow the water to cool, by giving heat to the outside air.

   i. Cooling System: Any system which exists to cool some part of a nuclear power plant.

   j. Coolant: Any liquid or gas which removes heat from an area.

3. There are several choices of options in nuclear power plant design:
   a. Fuel Type: natural Uranium; enriched with U-235; or Plutonium
   b. Moderator: water; graphite
   c. Coolant: air; water; liquid metal
   d. Velocity of neutrons: thermal; fast
   e. Use: power generation; breeders; research; atomic bombs

7.4
The Science of Meltdowns and Explosions

Introduction

One of the main concerns with nuclear power is the possibility of nuclear meltdowns and nuclear explosions. Images of Three Mile Island, Chernobyl, and Hiroshima are in the minds of many people. Let us then start looking at these issues more objectively.

Meltdowns

Overview

A meltdown exists when the fuel rods or the nuclear fuel pellets begin to melt inside the reactor core. A meltdown usually begins when there is some problem with the cooling mechanisms. The fuel rods are made of metal, usually a form of Zirconium. The fuel is usually Uranium, Plutonium, or a form of Uranium Oxide. When the reactor core becomes too hot, any of these items inside the core can melt.

There are several possible results of a meltdown:

1. Radiation is contained within the reactor.
2. Radiation leaks into the ground
3. Radiation leaks into underground water supplies
4. Explosion occurs

1. Radiation is Contained Within the Reactor

Most often, radiation is contained within the container. In this situation, no one is affected by radiation.

2. Radiation Leaks Into the Ground

If the heat causes the containment structure to crack, then it is possible for the radioactive isotopes to leak into the ground. In this situation, the liquid metal, the hot water, and all the radioactive isotopes will then seep into the ground below. As the water and liquid metal travel through the earth, these liquids carry the radioactive isotopes with them. As the liquids travel, they begin to cool. When the liquids are cool enough they will stop, and the radioisotopes will stop with them.

3. Radiation Leaks Into the Underground Water Supplies

As above, if the heat causes the containment structure to crack, then it is possible for the radioactive isotopes to leak into the ground. If there are underground water supplies nearby, then it is possible for radioactive isotopes to enter the underground water. This would be a bad situation because this water can lead to nearby wells and streams which supply drinking water.

4. Explosion Occurs

In a meltdown, more fission reactions take place than is desired. The numerous fission reactions will produce a great amount of energy. The release of a lot of energy, at the same time, is an explosion.

A second type of explosion, the Hydrogen explosion, can also occur. The hydrogen explosion is caused by hydrogen reacting with oxygen. (This will be discussed in greater detail in the section on explosions).

Note that it is possible for either type of explosion to occur inside the reactor and yet still be contained. This depends on the strength of the container and the size of the explosion.

Melting Points of Fuel Rods and Fuel Pellets

Overview

If the nuclear reactor is not kept cool enough then a meltdown can occur, where some of the fuel rods and fuel pellets change from solid to liquid state. In this form the pellets and rods produce much greater amounts of energy. The fuel is also more likely to migrate to undesired areas.

The first question to answer is: what exactly are the melting points of the fuel rods and fuel pellets? In order to answer that question, we must know the specific chemical involved. The first table below provides the melting points for most nuclear materials.

Melting Points of Fuel Rods and Fuel Pellets

Material	Melting Point, °C	Melting Point, °F
Plutonium	639 °C	1,183 °F
Uranium	1,132 °C	2,070 °F
UO_2	2,827 °C	5,121 °F
Zirconium	1,855 °C	3,371 °F
ZrO_2	2,677 °C	4,850 °F

Melting Points of Uranium Oxide Pellets

Many nuclear fuel pellets are actually Uranium Oxide rather than just elemental Uranium. The most common forms of Uranium Oxide in fuel pellets are UO_2 and U_2O_5. However, when making a fuel pellet of "Uranium Oxide" the pellet may contain any of the possible Uranium Oxides. There are seven possible forms: UO, UO_2, UO_3, U_2O_5, U_3O_7, U_3O_8, and U_4O_9.

The Uranium Oxide pellet can be a mixture of several forms of Uranium Oxide. Therefore, the melting point of a Uranium Oxide fuel pellet is not exact. The melting point of this type of fuel pellet will depend on the specific oxides, and the amount of each oxide, within each pellet.

Rather than trying to identify the exact melting point of every mixed oxide pellet, the melting point of a Uranium Oxide pellet is usually stated as a range. The official range for the possible melting points of Uranium Oxide pellets, as stated by the International Nuclear Safety Center, U. S. Department of Energy, is as follows:

Melting Point Range for Uranium Oxide Pellets
Range in Celsius:　　2,816.84 °C to 2,876.84 °C
Range in Fahrenheit: 5,102.32 °F to 5,210.32 °F

Nuclear Explosions

Overview

Explosions can occur as a result of two factors: 1) excessive chain reaction (uncontrolled fission/meltdown), or 2) hydrogen gas reacting with oxygen.

1. Explosion by Excessive Chain Reaction

If the fission reactions are uncontrolled then the chain reactions will produce a great amount of energy. Instead of just a controlled amount of energy being released throughout each day, we get all the energy from all the nuclear fuel being released at the same time. Such a great amount of energy, released at one time, can result in an explosion.

This explosion (excessive energy) is enough to burst a hole in some containers. Generally, if the wall is too thin then the energy released from the excessive chain reaction will create a hole. After the hole is created, radioactive isotopes can escape into air. A better designed reactor will have two containers. When the reactor has two enclosures, if the excessive chain reaction bursts a hole in the first container, they will not leave the reactor. The radioactive isotopes will be prevented from leaving due to the second enclosure.

2. Explosion by Hydrogen gas

Many fuel rods and some tubes in the reactor core are made of Zirconium. When water passes the Zirconium, sometimes a chemical reaction takes place. In this chemical reaction, Zirconium takes Oxygen from the water molecules, which leaves Hydrogen gas. Usually this is only a small amount of hydrogen created, and is vented away easily. This is usually not a problem.

However, if the Hydrogen gas builds then there is a potential for an explosion. Hydrogen gas reacts with Oxygen gas very explosively. Therefore, if there is enough Hydrogen gas in the reactor (as produced by the Zirconium-water reaction), and if there is enough Oxygen gas in the reactor then there will be a Hydrogen explosion.

Size of Potential Explosions in Nuclear Power Plant

An explosion due to excessive chain reaction will never be anything as large as an atomic bomb, such as at Hiroshima. The amount of fissionable material in a plant is far less than what is put into an atomic bomb. The fuel in a nuclear power plant has a maximum of 5% fissionable material, whereas the fuel used in a nuclear weapon has at least 90% fissionable material.

The bigger problem of an explosion at a power plant is that radioactive isotopes (by-products and fuel) are thrown into the air. This is what occurred at Chernobyl.

The other type of explosion is the hydrogen explosion. The size of a hydrogen explosion at a power plant can vary. The size of a hydrogen explosion will be approximately equivalent to an explosion that might happen at an oil refinery or in a chemical plant.

Summary of Meltdowns and Explosions

1. A meltdown exists when the fuel rods or the nuclear fuel pellets begin to melt inside the reactor core.

2. There are several possible results of a meltdown:
   a. Radiation is contained within the reactor
   b. Radiation leaks into the ground
   c. Radiation leaks into underground water supplies
   d. Explosion, contained or uncontained

3. There are two ways for explosions to occur:
   a. Excessive chain reaction (uncontrolled fission)
   b. Hydrogen gas

4. Explosions by excessive chain reaction:
   If the fission reactions are uncontrolled then the chain reactions will produce a great amount of energy. Such a great amount of energy, released at one time, can result in an explosion.

5. Explosions by Hydrogen and Oxygen:
   Many fuel rods and some tubes in the reactor core are made of Zirconium. When water passes the Zirconium the Zirconium takes Oxygen from the water molecules which leaves Hydrogen gas. Usually this is only a small amount of hydrogen gas and is vented away easily. However, if there is enough Hydrogen gas and enough Oxygen gas in the reactor, then there may be a Hydrogen explosion.

7.5
Three Mile Island

Introduction: Three Mile Island in Perspective

The incident at Three Mile Island provides as many positives as it does negatives. Because of this incident, improvements have been made in all areas of nuclear safety. This includes: structural safety designs, automatic safety features, emergency procedures, communication procedures, regulations, training, inspection, and so much more.

Three Mile Island was 25 years ago. In technological terms, that is a significant length of time. We should not think of Three Mile Island as proof that nuclear power is unsafe, but rather we should think of the incident as a starting point to see how far we have come.

When discussing Three Mile Island, we must put the dangers in perspective. The actual results of the incident are as follows:

1. All radiation was contained. No radiation escaped the plant.
2. No one was affected by the radiation. No one was harmed.
3. There were no fatalities.
4. The meltdown was stopped within 12 hours.
5. The operators followed procedures exactly as trained.
6. There were no fires.
7. There were no explosions.
8. The other reactor at the site never had any problems.

Overview of Events at Three Mile Island

The technical problems at Three Mile Island were primarily two faulty valves. These problems were resolved within a day. The lingering issue was the potential for a Hydrogen explosion.

We must remember that the story of Three Mile Island is more about uninformed fear and poor communication than about the actual nuclear hazards.

I have divided the story of Three Mile Island into the following stages:

A. Partial meltdown due to lack of water

B. Hydrogen gas

C. Reducing the amount of radioactivity should an explosion occur

D. The panic

E. The end results

Partial Meltdown due to Lack of Water

1. On March 28, 1978 routine maintenance was being performed on the water filters in reactor number two. At the time, the reactor was operating at 97% of capacity.

2. During this time, water became blocked by one of the filters in the secondary water loop. (The secondary loop is the heat exchange, in a BWR reactor with heat exchange.)

3. The potential problem: If water in the secondary loop was not continuing on its path, then there would be nothing to absorb the heat from the primary loop. If the primary loop could not give up its heat, that water would never cool, and the temperature inside the reactor core would keep rising. This continuous rise in temperature could create a meltdown of the reactor core.

4. The power plant system sensed the blockage. The plant automatically shut off the steam turbine. The plant automatically signaled the control room. This system operated just as it was supposed to.

5. The emergency control rods came down, as designed.

6. A valve in the primary loop opened, as designed, in order to let go of some of the steam.

7. What should have happened, but didn't: After a certain amount of steam was let go, the valve should have closed. However, the valve did not close. Furthermore, there was no automatic signal that would have let the operators know that the valve was stuck open.

8. Water (as steam) was now leaving the primary loop at a rate of 50,000 kg per hour (approximately 220 gallons per second).

9. Automatic auxiliary pumps started pumping water into the secondary loop (beyond the blockage point), just as was supposed to happen.

10. However, the valves to the secondary loop were stuck *closed*. None of the water being pumped was let inside.

11. At this point we have a valve stuck open in the primary loop, letting out lots of water. We also have a stuck valve on the secondary loop, which did not let in any additional water.

12. The operators were trained to check the water level in the reactor by looking at the water level in the primary loop. This value looked fine, but the operators did not know that steam was rapidly leaving, and they did not realize a valve was stuck open.

13. Of course, the steam pressure in the primary loop would eventually drop, and it did. The power plant had an automatic emergency system which would respond if the pressure dropped too low. This emergency system worked as designed, pumping in lots of water into the primary loop.

14. At this point the operators made a mistake.
   The operators saw the emergency system pumping water into the primary loop. What the operators should have done was to assume the system measured the pressure correctly, which would have told the operators that the primary loop must have had low pressure. That thought would have led them to the cause: the stuck valve.
   Instead, the operators assumed that the sensors were wrong, and that the primary loop was being *flooded* by this emergency response system. Therefore, the operators over-rode the system, and turned off the emergency water to the primary loop.

15. This was the situation for two hours, until the operators realized the problem of the stuck valve.

16. The operators realized that a valve was stuck in the primary loop, and fixed the problem. The total time from the initial water blockage to fixing the valve was 12 hours.

Hydrogen Gas

17. The meltdown problem was over, but there was a new problem: Hydrogen gas.

18. It was known that Hydrogen gas was being built up. The questions were: How much hydrogen? Would it dissipate or would it explode? Is the public in danger? Should we evacuate the area?

19. These were the questions that puzzled the experts. Information was changing and conflicting, which only added to the public's anxiety.

20. In fact, there was not enough oxygen in the core to cause a hydrogen explosion. However, at the time of the incident not all authorities were certain.

Reducing the Amount of Radioactivity should an explosion occur

21. If a hydrogen explosion occurred in the reactor core, not only would part of the plant be blown apart, but nuclear material inside the core (uranium and by-products) would be floating in the air. Assuming the worst case scenario (a hydrogen explosion) could occur, the authorities worked to reduce the amount of radioactive isotopes in the core.

22. The authorities reduced the radioactivity in the following way. First the operators vented the nuclear by-products into another building. Second, the operators diluted the water which contained the by-products. Finally, the operators sent the diluted nuclear by-product water into the local river.

The Panic

23. An emergency siren went off in the city. No one authorized that siren, and no one has ever admitted to turning the siren on. Yet the sounding of that siren added tension and fear among the community.

24. The threat of a Hydrogen explosion lasted for several days. There were conflicting reports on whether or not there would be an explosion.

25. Citizens and journalists did not understand the science. Two main facts were misunderstood: 1) If an explosion did occur, this explosion would likely be smaller than what the citizens feared. 2) The radioactive isotopes sent into the river were highly diluted.

26. No one was told to evacuate, but that made no difference to the citizens. 40,000 people voluntarily evacuated. They were determined to leave and they were very serious about leaving.

The End Results

There was no explosion. There were no fires. No radioactivity left the plant (except for the diluted radioisotopes which the authorities deliberately sent into the river). No one was harmed by radioactivity, ever, due to this event.

However, there was indeed a partial meltdown of the fuel rods in the reactor core. How much of meltdown occurred? The answer to that question requires some explanation. The inside of the reactor core could not be seen for a year after the event. During the year when no worker could enter, the reactor core remained hot. Therefore, while it is true that 40% of the reactor eventually melted (after a year), the amount of meltdown during the event was probably much less.

Summary of Three Mile Island

1. The technical problems at Three Mile Island were primarily two faulty valves. These problems were resolved within a day. The lingering issue was the potential for a Hydrogen explosion.

2. The story of Three Mile Island can be divided into the following stages:
   a. Partial meltdown due to lack of water
   b. Hydrogen gas, reducing the radiation if an explosion occurred
   c. The panic
   d. The end results

3. Partial meltdown due to lack of water

   Water became blocked by one of the filters in the secondary water loop. The power plant system sensed the blockage, and automatically made several actions to respond. All of these actions happened just as was supposed to. However, there were two stuck valves. One valve was stuck open, letting steam out, the other valve was stuck closed, not letting new water in. This was the situation for 2 hours, until the operators realized the problem of the stuck valve. The total time from the initial water blockage to fixing the valve was 12 hours. The problems related to the meltdown were over.

4. Hydrogen gas

   It was known that Hydrogen gas was being built up. The questions were how much, would it explode, and was the public in danger? Information was changing and conflicting, which only added to the public's anxiety. Assuming that a hydrogen explosion might occur, the authorities reduced the radioactivity by venting the nuclear by-products into another building, diluting that water, then sending the diluted nuclear by-product water into the local river. Later studies showed that there was not enough oxygen in the core to cause a hydrogen explosion.

5. The panic

There were conflicting reports on whether or not there would be an explosion. Media often exaggerated the facts. The details of nuclear power were commonly misunderstood. No one was told to evacuate, however 40,000 people voluntarily evacuated.

6. The end results

There was no explosion. There were no fires. No one was exposed to radiation. No one was harmed due to this event.

7.6
Chernobyl

Chernobyl: Overview and Perspective

Chernobyl was absolutely preventable. This was not an accident. The catastrophe created at Chernobyl was caused by incompetence and gross negligence.

The Chernobyl story in brief is as follows. Operators at Chernobyl performed a test. In order to run this test, they needed to reduce the power output. However, the operators reduced the power output by too much, and therefore had to increase the power again. That should not have been a problem if the operators raised the power output safely. However, the operators became more interested in performing the test than in safety issues, and therefore they made many bad decisions which violated safety procedures.

The operators violated safety procedures in three main ways: turning off safety features, over-riding safety features, and deliberately ignoring warning signals. All of these safety violations taken together created a catastrophe.

The most obvious safety features ignored were the control rods: at one point there were no control rods anywhere in the core. The computer gave a warning, which the operators promptly ignored. The operators started their test, and the explosions occurred.

In addition, the construction of the plant at Chernobyl was very different from any nuclear power plant in Western Europe or the United States. Even with all the operator negligence, many experts agree that if the plant was designed differently, then the troubles would have been contained.

Chernobyl: What Actually Happened

The disaster at Chernobyl was not an accident; the disaster was gross negligence. The result was a preventable catastrophe.

1. On April 25, 1986, Chernobyl reactor number 4 was scheduled for routine maintenance. While this was going on, the operators decided to run a test. Specifically, the operators decided to run a test on the steam turbine.

2. Some records have stated that this test was not authorized. However, similar tests had been performed before.

3. In order to run the test, the power output was gradually reduced.

4. The operators turned off the emergency cooling system. They did this so that the emergency system would not come on during the test.

5. The test procedure continued, lowering the power output even more. However, an error in their procedure reduced the power output to far lower than they wanted.

6. The operators wanted to increase the power, but they increased the power too quickly. They deliberately broke many safety rules in this process.

7. First, the operators raised all of the manual control rods away from the core. Raising all of the manual control rods certainly increased the power, but this act was in direct violation of safety procedures.

8. To prepare for the test, the operators decreased the steam pressure in the turbine. The decrease in steam pressure would normally trigger a safety feature which would shut down the reactor. However, the operators deliberately shut off this safety feature.

The operators reasoning was as follows: Due to the circumstances deliberately created for the test preparation, the steam pressure would drop. However, such a low steam pressure would automatically shut down the system. The operators didn't want the system to shut down because that would interfere with their test. Therefore, the operators blocked the signals from reaching the computer. By blocking the signals to the computer, the computer never knew that the steam pressure was too low.

In essence, the operators turned the signals off, which gave false readings to the computer. Therefore, the situation was unsafe, yet the emergency shut-down would never occur. Note that turning these signals off violated official safety procedures.

9. Another consequence of lower steam pressure was that the remaining control rods (the automatic control rods) were raised. At this point, all the control rods – manual and automatic – were raised.

10. For emphasis, let me remind you that at this point there were no control rods in the reactor. All of the control rods had been raised.

11. The computer gave a warning. An automated printout stated that the reactor should be shut down, as there was no means of controlling the reaction at this point. (Without any control rods, it is impossible to control the reaction.) The operators deliberately ignored this warning.

12. Let us pause and look at the situation before the actual test began: Steam pressure was lowered in order to run a test on the turbine. However, the water system is a continuous loop. The result was less water to cool the reactor. Furthermore, all the control rods (manual and automatic) were raised. At this point, there was no method of controlling the nuclear reaction.

13. The operators began the test. In less than a minute, the reactor started producing energy at a very rapid rate. The energy was produced at a much faster than rate it should (if the control rods were in place).

14. The operators tried to lower the manual control rods, but this took too long. By the time the control rods came down, the reactor core had heated up dramatically, the pressure inside the core (due to the numerous atoms created by all that fission) had increased dramatically, and the overall energy inside the core had increased dramatically.

15. The first explosion occurred: fission explosion. The energy released from the numerous fission reactions blew off the top of the reactor core.

16. The second explosion occurred: Hydrogen explosion. Hydrogen was created in the Zirconium–water reaction (see the earlier section on Hydrogen explosions). This Hydrogen reacted with the available oxygen from outside the reactor (remember that a hole had been blown in the top of the reactor), which therefore created a hydrogen explosion.

17. With large holes in the reactor, radioactive isotopes were able to leave the reactor core and travel through the air.

18. These explosions started more than 30 fires. One of these fires was close to reactor number 3, which was operating at the time.

19. Most of the fires were contained within a day. Yet the fire near reactor three was still burning, and this fire had to be handled carefully.

20. For four days the fire burned. For four days the radioactive isotopes escaped into the air. It was four days before the USSR asked for advice from other countries. With advice from Germany and France, the USSR dropped tons of limestone, lead, and Boron onto the site. Eventually the fire was put out.

Design Flaws in the Chernobyl Nuclear Power Plant

Before fearing a "Chernobyl" type explosion in the Unites States, we must note some differences between the plant at Chernobyl and plants in the United States. Many western nations had forbidden the Chernobyl plant design long before the disaster. These nations had refused to allow this design due to a variety of safety concerns. The United States never considered this design, and no plants in the U.S. use this design.

The Chernobyl plant did not have significant outer protection. In contrast, all plants in the Unites States have a strong outer encasement. Nuclear reactors in the Unites States have an outer layer of cement, which encases the reactor core and the cooling systems. In contrast, the plant at Chernobyl did not have this. Many experts agree that if this outer protection existed, then the first explosion would have been contained.

Furthermore, this reactor used graphite as the moderators. When graphite is used as the moderator, it tends to get hotter than water does. Note that there are some plants in the United States which use graphite as the moderator.

Summary of Chernobyl

1. The disaster at Chernobyl was not an accident. What happened at Chernobyl could only be called gross negligence.

2. The operators decided to run a test. In preparation for this test, the operators ignored many safety procedures, including:

    a. The operators turned off the emergency cooling system.

    b. The operators raised all of the manual control rods out of the core.

    c. The operators deliberately shut off a safety feature which would shut down the reactor if steam pressure was too low.

    d. The computer gave a warning that the reactor should be shut down, but the operators deliberately ignored this warning.

3. When the operators started their test, the reaction got out of control, and the explosions occurred. It was four days before the USSR asked for advice from other countries. Radioactive isotopes were allowed to vent into the air, and fires were allowed to burn, for a total of four days.

4. The Chernobyl plant design was only used in the USSR. Other nations thought the design to be unsafe before the explosion occurred.

5. The Chernobyl plant did not have significant outer protection. In contrast, all plants in the Unites States have a strong outer encasement. Many experts agree that if this outer protection existed, then the first explosion would have been contained.

7.7

The Nuclear Accident in Fukushima, Japan

Introduction

The nuclear accident in Fukushima, Japan is the most recent of nuclear accidents. The major cause of the disaster was the earthquake, in combination with the tsunami that followed. The secondary cause was venting too much hydrogen into the air at one time, causing the first explosion.

Only recently has the history of this tragedy been analyzed. In this chapter we will look at some of the discoveries made so far.

What Happened in Japan

1. The Primary Cause of the Accident was the Earthquake

On March 11 of 2011, Japan experienced a 9.0 earthquake, one of the largest earthquakes in recent history. Although the designers of the facility did earthquake studies prior to construction, their level of design was no match for this magnitude of earthquake.

There were six nuclear reactors at the facility. Reactors #5 and #6 were down for maintenance before the earthquake, reactor #4 was cooling but not operational, and reactors #1, #2, and #3 were operational at the time.

The most significant damage (in terms of creating the disaster) was damage to the internal power infrastructure. Although the reactor cores continued to produce energy, the electrical and mechanical infrastructures within the power plant were damaged. This meant that normal operations such as computer systems, remote switching operations, and water pumps could not continue because they had no power source.

Note that without such power, the water pumps could not circulate water around the reactor cores to cool the reactor. If water could not be provided to the reactors soon, the reactors would produce so much energy that a meltdown or an explosion was inevitable.

2. Back-up Generators Started up, as Designed

The designers of the power plant had planned for such a contingency. A series of back-up generators, mostly diesel, were immediately put into use. As designed, these back-up generators would provide power for the entire operation of the nuclear power plant. As designed, the back-up generators provided power to the water pumps. The water continued to circulate, and the reactors would soon cool to normal temperatures. This situation would have been acceptable, if not for the tsunami.

3. The Tsunami hit Japan, Destroying the Back-up Generators

After with being hit by a powerful earthquake, Japan was also hit by a large tsunami. The facility did have a 20 foot seawall, but the 40-50 foot tsunami wave easily overshot this seawall and swamped the back-up generators, making them absolutely useless.

There were two other back-up generators on the hill, which were not affected by the tsunami wave. However, the switching mechanism was also swamped, making those generators essentially useless.

Without any power, the water pumps could not operate. At this point, the reactor cores continued to produce energy without any means of cooling. A meltdown or explosion would be inevitable.

4. Zirconium Fuel Rods Produce Hydrogen

As in the other nuclear incidents, the Zirconium fuel rods continued to produce Hydrogen, without any means for controlling or venting. The Hydrogen in the reactor core builds and builds.

5. Decision to Vent or Not to Vent

As in Three Mile Island, the engineers discussed the Hydrogen situation. Was there enough to cause an explosion? Can we vent it safely? The decision was made to vent the Hydrogen.

6. Venting Hydrogen Causes the First Explosion

Remember that one of the ways a nuclear explosion can occur is through Hydrogen reacting with Oxygen. When a large enough supply of Hydrogen interacts with a large supply of Oxygen (as exists in the air), the result can be an explosion.

The engineers decided to vent Hydrogen into the air, starting with units #1 and #3. However, the amount of Hydrogen being vented was so great that there was a large explosion in #3. This was the first explosion at the nuclear power plant. At the same time, units #2 and #4 were beginning to seriously overheat.

7. Other Reactors at the Power Plant quickly Explode and Burn

Consider the other reactors at the time. Because no reactor had a cooling mechanism, each reactor was building energy at an alarming rate. This would cause meltdown or explosion from the inside. Similarly, each reactor was building up Hydrogen, which if coming in contact with air could explode as the first reactor did. Furthermore, add heat to any flammable environment and you will create a fire. Therefore, after the first reactor exploded, the other three quickly exploded as well. One by one: #3, then #2, then a fire in #4, then another explosion in #3.

8. Radioactive Materials Spread through the Air

As in the case of Chernobyl, once the reactors were blown open the radioactive material was emitted into the air. At this point, the reactors are releasing enormous amount of radioactive isotopes into the air, to be floating around any number of miles.

This is the most serious consequence. Those radioactive isotopes can be inhaled, breathed into the body and absorbed into the blood. These isotopes can also enter the water supply or the soil of farms, in which case these isotopes will be ingested. Once inhaled or ingested the radioisotopes will stay in the body for a long time. Most significantly, the radioactive decay will damage the person's body from the inside, where it is easiest to do the most damage.

9. Japan is still Working with the Earthquake Aftermath in Other Regions

During this time, the rest of Japan was still dealing with the consequences of the earthquake. The infrastructure was damaged throughout Japan, including roads, buildings, and water supplies. In addition, response teams and emergency responders were already working 24/7. Therefore, getting assistance to the damaged nuclear power plant was difficult. For the same reasons, evacuating the citizens from the area was also very inefficient.

10. Using Sea Water to Put out the Fires

Japan is an island, surrounded by the Pacific Ocean. At first glance it makes sense to bring in vast quantities of water from the ocean to put out these fires. There is only one problem: sea water will corrode metals and other materials. Therefore, as the fires were being put out with vast amounts of sea water, this same water also corroded many of the parts within the power plant.

11. Generators Working, but Pumps Still Not Working

Eventually the power plant gets some back-up generators working. Some of the power plant components are operational. However, some of the components are corroded beyond use. Certain valves and water pumps in particular are barely operational or are totally useless. This means that the internal water system, normally used to cool the reactors, is still of minimal use.

12. Fires are Put Out, the Reactor is Contained

Eventually the fires are put out. Volunteers arrive to contain the radiation leaks, and contain the nuclear reactor altogether.

Lessons Learned

The nuclear power plant incident in Japan is still being studied and evaluated. Lessons are yet to come.

However, one early concept has been discussed: understanding your surroundings. Earthquakes and tsunamis have hit Japan before, and the Japanese know this. Therefore it is often asked why the Japanese designers didn't consider their environment when building their plant.

Indeed, investigation from Japanese officials has shown that the designers of the facility did NOT adequately prepare for the magnitude of earthquakes or tsunamis which were possible. It was stated that the energy company had "overconfidence" in their facility, particularly when compared to the magnitude of well-known natural disasters throughout Japan's history.

The incident at Japan reminds us that we must always consider our surroundings and potential for natural disasters when building any energy technology system.

7.8
Making Nuclear Power Plants Safer

Introduction

Overall, nuclear power plants are relatively safe. Consider the number of nuclear power plants operating throughout the world over the last 30 years to the few number of incidents, and you will see that the nuclear power industry has a relatively high rate of safety.

However, when a nuclear incident does occur, the results can be serious – very serious indeed. Therefore, we must have a zero tolerance policy regarding nuclear accidents. In order to achieve this we must implement many specific solutions, in several areas of nuclear power plant safety.

In this chapter, we will look at some of the more important methods for making nuclear power plants safe. Many of these tips come from the American Nuclear Society, the Nuclear Regulatory Commission, or professional nuclear engineers.

Consider Local Environment and Weather

It is important to consider the local environment and weather before choosing the final site. Similarly, all designs for construction and emergency planning must take into account regional environmental factors. The greatest environmental factors which can damage a nuclear power plant are: earthquakes, tornadoes, hurricanes, and flash floods.

In general, earthquakes cause the greatest damage. Shaking the structure can cause cracks and breaks anywhere in the structure. These cracks may not be visible, and the engineers may not be aware of them. In addition, these cracks will grow over time: with the stresses of water pressure, steam pressure, heat, and repeated earthquakes, these initial cracks will eventually cause large holes. Of course, earthquakes of major size, 7.0 or larger, are likely to cause major holes immediately. Everything from piping systems to the outer containment structure may be vulnerable to this shaking. Therefore it is not recommended to build nuclear power plants in areas which are prone to major earthquakes.

Tornadoes and hurricanes are the next major environmental factors. The winds in tornadoes and hurricanes can thrust objects into the power plants with enough force to penetrate the outside. Therefore, nuclear power plants should not be placed in areas which have frequent tornadoes and hurricanes. In addition, if there is a possibility of a tornado or hurricane in the area, the facility is best placed underground, with only the cooling towers exposed. If this is not an option then the containment structure must have an extra layer which can withstand the greatest wind force naturally possible.

Flash floods and tsunamis can be a factor. The flood is not likely to damage the main structure of the power plant, however the sudden amount of water can damage the surrounding equipment, including power lines and generators. The water can also sink into the ground to such a level as to make the soil below the power plant unstable. If flash floods or tsunamis are likely, it is best to design channels, drainage ditches, and reservoirs around the power plant, which will diver the water away from the reactor.

Most other environmental factors are minor, and do not affect the structure of the plant. The structure of most power plants will be able to withstand the high winds of most storms, heavy rain, snowstorms, hail, and regular weathering.

Design and Construction of Nuclear Power Plants

The design of the power plant is the first area of safety. The design and construction are the first assurances against a nuclear power plant failure. When a nuclear power plant is designed and constructed, this power plant should be built with the following:

1. Adequate Number of Control Rods

2. Adequate Number of Emergency Control Rods

3. Moderators

4. Emergency core cooling system

5. Emergency water supplies at various points along the water loop

6. Gauges, lights, and other automated warning systems

7. Sensors which trigger automatic emergency systems

8. Automatic shut-down of reactor core at maximum temperature or at maximum Hydrogen level

9. Two-layer containment structure around the core

10. Steel containment below the core (a giant steel tub) to contain a meltdown

The specific dimensions, amounts, and materials for each of these will depend on the specific size of the power plant. However, these are the essential elements to look for when reviewing the design of a power plant. It is also important to ask the designers questions on these topics, such as: "How many of [each item]" do you have?", "Where are they placed? and "Why is this the best material/best placement/right amount for adequate safety?"

Remember the Titanic: the disaster could have been averted if the captain had learned of the iceberg soon enough. More people could have survived if there were enough life boats. Ask the questions now, in the design phase, so the plant operators are ready to minimize damage should an incident occur.

Control Room

The control room of any large facility can be very complex. The operators must be given all the information they need, and yet the information must be presented as simply as possible. When an emergency arises, an enormous amount of important data can come very quickly, and a properly designed control room can be the pivotal factor in preventing a crisis. Tips for designing better control rooms include the following:

1. Place computers showing the same information in different areas of the plant so that several employees can see the same information.

2. Install multiple video cameras throughout the power plant so operators can actually view the events inside the reactor. These visuals can be compared with the information being presented by the computer.

3. Install computer programs that can quickly analyze the combinations of variables, and warn the operators of potential problems. Computers must also have user-friendly menus and fast response times.

4. Periodically replace the computer systems with the latest technology, particularly with the greatest memory and the highest speed.

Training of Operators

Proper training of nuclear power plant operators is essential to the safe operation of the plant. Nuclear power plant operators must be held to a higher standard than the operators at other types of facilities.

These operators must have a detailed understanding of the science of nuclear power, as well as a detailed understanding of the specific technologies unique to their power plant. Operators must also be masters of every possible contingency.

Tips for training nuclear power plant operators include the following:

1. All nuclear power plant operators must go through rigorous training, using a power plant simulator.

2. The federal government should invest in building several varieties of nuclear power plant simulators because nuclear power plants are not identical.

3. Training schools for nuclear power employees must have strict standards.

4. A significant part of the course must be on dealing with various emergencies.

5. In order to be certified, a nuclear power operator must pass a test which simulates a real nuclear accident.

6. All nuclear power plant operators must practice regularly for emergency situations. Operators must practice in a simulator every five years in order to remain certified.

7. Operators being trained or recertified must be subjected to various emergency simulations, not just one. A power plant is complex, many individual events can occur, and therefore operators must be prepared for any number of possible situations.

Regulations and Government Inspections

The Federal and State governments must take an active role in ensuring the safety of nuclear plants. Tips for regulations and inspections include:

1. The design of every proposed nuclear power plant must go through rigorous checks by state and federal regulators before being approved.

2. The construction of the nuclear power plant must be inspected regularly as it is being built.

3. Nuclear power plants must be inspected every few years by federal and state inspectors.

4. Plant managers should invite government inspectors to observe during down times (such as maintenance as refueling).

Maintenance

Regular maintenance is essential. All structures will become weak with age, and all mechanics will wear down with use. Tips for maintenance include the following:

1. <u>All Parts of the Plant Must be Maintained on a Regular Schedule</u>
Every filter, every valve, every rod, and every sensor must be checked regularly. Each part must be cleaned, repaired, or replaced as necessary.

2. <u>Power Must be Reduced Before Any Maintenance Occurs</u>
The power output of the nuclear plant must be reduced, ideally to zero power, before any maintenance is performed. This will prevent any accident from occurring during maintenance.

3. Emergency Equipment Must be Inspected Every 1-5 years

Emergency equipment must be inspected and tested periodically. An ideal schedule is to test 20% of the emergency equipment each year. This schedule will test all emergency equipment over a period of five years, and also makes sure that at least some emergency equipment is fully functioning every year.

Communication and Emergency Planning

If a problem does occur, the dangers can be kept at a minimum and the problem can be fixed most easily if we have effective communication systems and response planning. Communication procedures must be put into writing, and these procedures must be practiced. Communication procedures should be written down for the following areas:

1. Chain of command for the nuclear power plant
2. Notifying the government
3. Local emergency response procedures
4. Updating the public
5. Evacuation criteria
6. Evacuation procedures
7. Coordination with community groups

These communication procedures must be practiced annually for the following organizations: the nuclear power plant, the local government, local emergency response teams, and international nuclear response teams. Each of these groups is operated independently, yet each must have its own system for emergency response and communication. Each group must also practice those systems regularly, including coordinating with the others.

Defense in Depth

The nuclear power industry has always designed with a concept called "Defense in Depth." The Defense in Depth philosophy is simply the design of redundant systems. The idea is that if one device of the power plant system fails to work, there will be another device ready to function. Every nuclear power plant built in the United States must be designed with "Defense in Depth" back-up systems.

When considering a design for a nuclear power plant, there are always two questions regarding Defense in Depth: How much redundancy is needed? Where should these redundant devices be installed? There are no definitive answers to these questions. Nuclear engineers and nuclear activists have constant discussions on the matter.

Nuclear Safety Standards

There are two excellent sources for Nuclear Safety Standards: the American Nuclear Society, and the Nuclear Regulatory Commission. I have listed a few of the most important fact sheets and regulations from these organizations in the Appendix.

Chapter Summary

1. Nuclear power plants do have hazards. Although the hazards are not likely, the results can be serious. Therefore, we must have a zero tolerance policy regarding nuclear accidents.

2. Consider the local environment and weather when choosing a site and designing for safety. Earthquakes, tornadoes, and hurricanes are the most serious environmental forces.

3. Design a nuclear power plant with numerous safety features, including control rods, moderators, cooling systems, sensors, containment systems, and automatic shut down systems. Make sure that there are adequate amounts of these safety features, that they are made of the best materials, and that they are properly placed

4. Design the control room to provide the necessary information, yet also to be easy to use. Video cameras and redundant systems are particularly important.

5. Proper training of nuclear power plant operators is essential. Nuclear power plant operators must be held to a higher standard than the operators at other types of facilities. Operators must have a detailed understanding of the specific technologies unique to their power plant and be masters of every possible contingency. They must also be rigorously trained and practice emergency response frequently.

6. The Federal and State governments must take an active role in ensuring the safety of nuclear plants, in design, construction, and maintenance.

7. Regular maintenance is essential to a safe power plant. Every valve, pump, switch, rod, and room must be inspected frequently.

8. Power plants must be reduced to a very low level of energy output before any maintenance will be done.

9. Communication procedures and emergency response procedures must be written and practiced annually for the following organizations: the nuclear power plant, the local government, local emergency response teams, and international nuclear response teams.

10. Nuclear power plants must have several redundant safety systems, in case one fails. This is known as "defense in depth".

11. The American Nuclear Society and the Nuclear Regulatory Commission have numerous fact sheets, reports, and best practices for nuclear power plant safety.

7.9
By-Products and Radioactivity

Introduction

Remember that the products of the fission reaction are the things we want. These products are energy and neutrons. Anything else is considered a by-product. Thus, the by-products in nuclear power are all the smaller atoms created by the splitting of Uranium.

Normally we would just consider these by-products as trash, and pay no attention. However, we pay attention to this trash because many of these by-products are radioactive, and radioactivity can affect the health of humans.

Radioactivity Basics

Overview

As discussed earlier, radiation exists because an atom which is unstable will want to become stable. The atom will become stable by emitting something from the nucleus. The item emitted is the decay. This also changes the identity of the element.

Also remember that radioactivity is a natural process. There are over a hundred isotopes which naturally undergo radioactive decay. If we never build another nuclear power plant there will still be radiation from these natural sources.

Types of Radioactivity: Alpha, Beta, and Gamma

Name	Symbol	What it is	Shielding material
Alpha particle	α	Helium atom	paper, clothing, skin
Beta particle	β	Electron	wood, plastics, thin metal
Gamma ray	γ	electromagnetic energy	lead, concrete

The Net Effect of Alpha decay is:

1. Helium atom is jettisoned from the nucleus.
2. Atomic # decreases by two. Thus the identity of the element changes to the *left* by *two* elements on the periodic table.
3. The isotope # decreases by four.

The Net Effect of Beta Decay is:

1. Electron is jettisoned from the nucleus.
2. Atomic # increases by one. Thus the identity of the element changes to the *right* by *one* element on the periodic table.
3. Isotope # remains the same.

The Net Effect of Gamma Decay is:

1. A burst of a short electromagnetic wave is given off.
2. Gamma affects the atom by changing its energy.
3. Gamma radiation does *not* change the identity of the atom.
4. Gamma does not occur by itself. It must accompany alpha or beta.
5. Gamma does not always happen. Even if alpha or beta radiation does exist, gamma does not have to exist.

Basic Process for each type of Radioactive Decay

Alpha Decay (Helium atom)

   Alpha decay is the loss of a Helium atom. (Two protons and four neutrons leave the nucleus together, as one Helium atom.) The net effect is that the type of atom has changed. It is now two atomic numbers less than the original atom. We are now two atoms to the left on the periodic table.

>   Example #1: Alpha Decay
>   Equation: Lead(82)–204 → Mercury(80)–200 + He
>   Description: Lead (atomic #82), becomes Mercury (atomic #80). You will also notice that the isotope number decreases by four.

>   Example #2: Alpha Decay, with Gamma
>   Equation: Radium(88)–226 → Radon(86)–222 + He + gamma ray
>   Description: Radium (atomic #88) becomes Radon (atomic #86). Helium atom and gamma radiation are emitted.

Beta Decay (electron): Details of the Process

Beta decay is the loss of an electron. This electron comes from a neutron in the nucleus. The process works as follows: A neutron is merely the term for a particle where a proton and electron are joined. Therefore, when beta decay occurs, one of the neutrons (proton + electron) gives up its electron. This leaves just a proton.

After a beta decay occurs, we have one "more" proton. Recall that the identity of an atom is determined solely by the number of protons. Therefore the identity of the atom changes, and is now one element to the right on the periodic table.

Remember also that we didn't really add a proton. That proton was already there, it was just previously a neutron. We merely took away an electron from a neutron, which resulted in a proton.

Example #3: Beta Decay
Equation: Copper(29)-66 → Zinc(30)-66 + electron
Description: Copper (atomic #29) becomes Zinc (atomic #30).

Example #4: Beta Decay with Gamma Decay
Equation: Potassium-40 → Calcium-40 + electron + gamma
Description: Potassium becomes Calcium (atomic #20).
Electron and gamma radiation are emitted.

Gamma (electromagnetic wave): Details of the process

Gamma radiation is a burst of electromagnetic energy. The energy of a single burst of gamma radiation can vary, depending mostly on the particular source. The energy of a single burst of gamma radiation can vary from about 100 eV to 10 million eV. The most common value is approximately 1 Million eV.

If we are to take one value to work with, we should either take the highest energy value, since it would be the most dangerous gamma burst for the longest distance, or the most common value, since that is what we would be likely to encounter. Therefore:

1. Highest energy of gamma radiation $=10$ MeV$=1.6 \times 10^{-12}$ Joules
2. Most common value of gamma radiation$=1$ MeV$= 1.6 \times 10^{-13}$ Joules

Remember that gamma radiation must accompany alpha or beta radiation. Also remember that gamma radiation does not necessarily come, even if alpha or beta radiation does.

Many By-Products of Fission

Overview

When we discussed fission, we used the primary example of Uranium splitting into Krypton and Barium. This was indeed the first example discovered, and is still the most common result of Uranium fission. However, there are many more possible combinations.

Remember that as long as the number of protons of the by-products equals 92 (the number in Uranium), many combinations of smaller particles are possible. Some of the most common combinations of smaller atoms from a Uranium split include: Krypton and Barium; Tin and Molybdenum; Tellurium and Zirconium; Lanthanum and Bromine; and Xenon and Strontium. However, again, many combinations of by-products are possible.

By-Products Themselves can Do Many Things

In addition to the initial creation of by-products from fission, the by-products themselves can do many things. By-products from fission can:

1. Absorb neutrons
2. Fission (flying neutrons create fissions in these by-products)
3. Decay, if the by-products are radioactive
4. Become radioactive, if the added neutron makes them radioactive

This means that in addition to the variety of isotopes created by the Uranium fissions, these isotopes create their own isotopes from their own processes. After the fission process has been going on for some time, we may have quite a range of isotopes in the mix. Some engineers have estimated that a nuclear reactor can get 100-300 possible isotopes in the mix of a nuclear reactor after it has been operating for several months.

Which radioactive isotopes are created as a result of nuclear power? We don't know for certain at this time. Scientists must do more studies in order to answer that question.

Hazards of the Many By-Products

What are the hazards of many by-products? The hazards depend on what isotopes are in the mix, and this mix of isotopes was created from the variety of natural processes mentioned above. Some isotopes are stable and are perfectly harmless. Other isotopes are radioactive and will decay. Of those isotopes that decay, some will disappear in seconds while others will last a few years. Scientists must do more studies in order to know the specific hazards from each nuclear power plant.

Half-Lives: Radioactivity and Time

Overview

Knowing that an isotope is radioactive does not tell us the length of time the isotope will remain radioactive. We know that the nucleus will change, but the time required to change can be anything from seconds to millions of years.

Furthermore, the rate of decay for a radioisotope is not a constant rate. Unlike most natural processes, the rate of radioactive decay cannot be measured at a constant rate. We cannot say, for example, that an isotope will emit 10 beta particles per second. However, we do know that a certain amount of radiation will occur in a specific period of time.

Understanding Decay in Practical Terms

To get a practical understanding of half-lives and radioactive decay, let us begin with an analogy: leaves falling off of a tree. Suppose that we know that all 50 leaves of a particular tree will fall to the ground within two months. We know from experience that the tree will always shed 50 leaves within two months. However, we do not always know how fast the leaves will fall within those two months.

For example, the leaves could fall:

    a. 1 leaf a day for 50 days
    b. 10 leaves fall every other day.
    c. 1 leaf a day for 20 days, then 30 leaves fall all at once
    d. Or any other variation of bursts over the two months

The combinations of bursts and no bursts are almost endless. The leaves can fall just a few at a time, or there can be a few bursts followed by days of nothing. We can know that, in the end, 50 leaves will fall with two months. However, we don't know how the leaves will fall within those two months.

The decay of a radioactive isotope is similar. As with the analogy of the leaves, we know that a certain amount of decay will occur within a certain time, but we cannot pinpoint it down any more than that. For example, we might know that an isotope will emit 1,000 beta particles in a month. However, we don't know how fast that decay will occur within that month. In a similar way to the analogy of the leaves above, radioactive decay could occur as a few particles per day, or the decay could be a burst of many particles followed by days of no activity. We can never know for sure.

The best that we can pinpoint the rate of decay is a measurement called the "half-life." The half-life is the time required by the radioactive isotope to decay to half of its original amount. A shorter half-life means that: more decay (more emissions) are likely to occur in a given amount of time, and the total amount of radioactive isotope will be gone quickly. A longer half-life generally means that fewer decay will happen in a given amount of time, and the total amount of radioactive isotope will be around for many years. For data on the half-life of each radioactive isotope, see the Appendix.

Relative Hazard of Various Half-Life Times

Note that a longer decay time does not necessarily mean the isotope is more dangerous. In fact, these isotopes might be safer. Consider an isotope with a half-life of millions of years. Although we cannot know when bursts will happen, they are generally not frequent. You might be able to stand on that spot for several months and yet no decay occurs; you might be gone when the next burst occurs. Of course, you could be at that spot when a large burst occurs. However, with a longer half-life the odds are in your favor.

On the other hand, if you are near a radioactive isotope with a short half-life, the odds are that you will be there when a burst happens. Yet, these isotopes will become totally safe in a short time.

Summaries of Radioactivity and Half-Life

<u>Summary for Process of Radioactivity</u>
1. Radiation is a natural process.

2. Radioactivity is the process of an unstable atom transforming into a stable atom.

3. The basic process for transforming one element to another is the same for all types of radiation: Something is jettisoned from the nucleus. This is the decay. The result is a change in the nucleus. A change in the nucleus means a change in the identity of the element.

4. There are three types of Radioactivity: Alpha (helium atom), Beta (electron from the nucleus), and Gamma (electromagnetic energy).

5. The net effect of Alpha decay is:
   a. Helium atom is jettisoned from the nucleus.
   b. Atomic # decreases by two.
   c. The isotope # decreases by four.

6. The net effect of Beta Decay is from the nucleus:
   a. electron is jettisoned from the nucleus.
   b. Atomic # increases by one.
   c. Isotope # remains the same.

7. The net effect of Gamma Decay is from the nucleus:
   a. A burst of a short electromagnetic wave is given off.
   b. Gamma affects the atom by changing its energy.
   c. Gamma radiation does not change the identity of the atom.
   d. Gamma does not occur by itself. It must accompany alpha or beta.

<u>Summary of By-Products</u>
1. There are many possible fission products. As long as the number of protons of the by-products equals 92 (the number in Uranium), many combinations of smaller particles are possible.

2. The hazards of isotopes in a nuclear reactor depend on what isotopes are in the mix. This mix of isotopes is created from a variety of natural processes. Scientists must do more studies in order to know the specific hazards from each nuclear power plant.

Summary of Half-Life
1. The half-life of a radioactive isotope is the time required for half of the sample to decay.

2. The rate of decay is not exactly constant. The decay could occur in large bursts followed by a long time of no decay.

3. If the half-life is one hour, we do know how many atoms will decay within that hour, but we do not know how many atoms will decay per minute or per second within that hour.

4. A short half-life means that the sample will decay quickly. If you isolate the sample then you can return shortly and most of the atoms will be totally safe.

5. A long half-life means that the sample will decay slowly. It is possible to stand near the sample for a long time without a single decay occurring. If the sample does decay in your presence, the decay will probably be a very small amount.

6. If you stand near a sample with a short half-life, you will probably be exposed to a large amount of decay. This is because many bursts of radioactivity occur in a short time.

7. If you stand near a sample with long-half life, you will probably be exposed only to a very small amount of decay. This is because the bursts of decay occur less frequently.

8. The fact that a sample is radioactive for millions of years does not mean that it will be emitting decay regularly all that time. Days, weeks, or months may pass before the sample emits any decay again.

7.10
Health Issues of Radioactive Decay

Introduction

Introduction

There are many factors which determine just how harmful a nuclear by-product might be. In this chapter we will look at all of the factors of radioactivity as related to human health. Note that this chapter assumes that radioactive decay is in the air or in the water at the start. However, in reality radioactive decay is rarely released into the environment.

Mechanisms of Each Decay Type in Damaging Body

All three types of radioactive decay have the same net effect on humans: ionizing the cells. However, each type of radioactive decay does this in different ways.

The alpha and beta radiations are particles. Each time they hit a cell molecule they ionize it. This is much like a pinball bouncing around inside your body. These particles go bouncing around, hitting cells, and ionizing each molecule that the particle hits. The amount of cells that one particle hits depends on the energy of the particle.

In contrast, gamma is a burst of energy. Gamma is more like a knife; it does one very significant cut. Gamma radiation hits one molecule or one cell, then the damage is done. There is no energy leftover. Gamma radiation is highly localized, but it is certain to cause significant damage to any cell that it hits.

Primary Effects of Radioactive Decay on Health

Introduction

There are many possible health effects from too much radioactive decay entering the body. When a cell becomes damaged several things may happen:

1. The cell repairs itself completely.

2. The cell stays permanently damaged, but does not affect the other healthy cells. The extent of the health problem depends on which cells were damaged, and how many.

3. The DNA of the cell becomes damaged, in which case cancer is created. (See below).

Cancer Versus Localized Damage

Most of the time when a cell becomes damaged, that damage is localized. It is a damaged cell, but everything in the rest of the body is fine. However, suppose that the same radiation hits the cells in such a way to not just damage each cell, but to damage the DNA. Now we have a problem. The "building code" for the cell has changed. As the DNA replicates to form new cells, the cells are created with this new set of bad building codes, and therefore new faulty cells are created. This is known as a cancer.

The best way to stop a cancer is to destroy all the bad building codes. This usually means deliberately killing off the cancerous cells. Unfortunately, it is often difficult to destroy the bad cells without harming some good cells at the same time.

Radiation Burns (beta)

Beta particles tend to cause skin damage commonly referred to as radiation burns. Beta particles are small enough to enter the skin. These particles damage the skin, causing lesions that look similar to a traditional burn. However, the good news is that the skin has absorbed the beta particles. The beta particles lose all their energy in the skin cells, and therefore the beta particles do not reach any interior of the body. Beta particles rarely go beyond the skin to the internal organs.

Overall Health Dangers of Alpha, Beta, and Gamma

In this section we will look at each type of radiation and summarize the overall hazards.

Alpha decay releases Helium atoms. Natural helium is not really dangerous. However, alpha particles at very high speeds can ionize human cells. The likelihood of Helium entering your body is small. Helium atoms can be stopped by protection as thin as a piece of paper or regular clothing. The only way for an alpha particle to get into the body would be to inhale it.

Beta decay is the release of electrons from the nucleus. Electrons from beta decay harm us by ionizing the cells. Skin lesions, called radiation burns, are the most common result of exposure to beta decay. Radiation damage from beta decay is usually limited to the surface of the body. Beta particles rarely go beyond the skin into the internal organs.

Beta decay can enter your body relatively easily, and they can ionize cells easily. However, they do not travel far. Electrons, even with their small size, tend to go only a limited distance in a human body (at most only a few millimeters). In air, beta particles will travel only a few feet. A relatively thin protection, such as certain plastics and thin metals, can prevent beta particles from reaching a person.

Gamma decay is a small, highly energetic wave. Because gamma is small, it can enter your body easily. Gamma decay can reach the internal organs. Because gamma is highly energetic, it can do much more damage to your cells than beta or alpha. Gamma decay is the main form of radiation that causes cancer. Therefore, gamma decay can be the most hazardous of all forms of radiation.

The energy of gamma radiation will spread out over distance. The further the gamma burst travels, the less energy at any point in the spread. Eventually, the energy from gamma will be so small as to not affect humans. Gamma decay can only be stopped by lead, concrete, or large tanks of water.

Routes of Entry into the Body

Introduction

There are four possible ways for radiation to enter your body. However, not all of these routes are equally likely to occur. These four routes are:

1. Traveling through the skin
2. Inhalation
3. Ingestion – in the drinking water
4. Ingestion – in the food

Radiation Entering Through the Skin

Of all methods of exposure to radiation, skin contact is the most likely. Furthermore, when engineers discuss shielding from radiation, they are primarily preventing radiation from traveling through the skin.

Alpha decay (Helium) is too large to enter the skin. Only the most energetic alpha particles will penetrate the skin, and even then the particles will only penetrate a few micrometers. Alpha decay will be stopped by regular clothing, paper, and in most cases, by the skin itself.

Beta can enter the skin but will not reach the internal organs. Thin shielding such as certain plastics will shield you from beta. Gamma can penetrate the skin very easily, and in some cases can reach the internal organs. A thick shield of lead is required to stop gamma.

Inhalation of Radiation

Regarding the inhalation of radioactive decay, the primary decay to consider is alpha. Since an alpha particle is too large to enter the skin, the only way for alpha to enter the body is through inhalation. However, remember that alpha is Helium, and many people have inhaled small amounts of Helium without any consequences.

Most of the time there will be no other radiation than alpha particles to inhale. However, if there is an explosion which is not contained, as happened in Chernobyl, then inhalation is a greater issue. The problem with a case like Chernobyl is not just radioactive decay, but the radioactive isotopes themselves.

If the radioactive isotopes are in the air, then a person could inhale these radioisotopes. The radioactive isotopes in the lungs can attach to the oxygen, which is then absorbed in the blood. Once in the blood, the radioisotopes can be spread to any part of the body. In such a case, radioactive decay will occur directly from the inside of the body for a period of time (until the body gets rid of those radioisotopes.)

Ingestion of Radiation Through the Drinking Water

If radioactive decay is in the drinking water then it will not last long. Alpha particles will be able to stay in the water the longest. However, as these are Helium atoms, these particles will tend to rise to the surface of the water and escape as Helium gas into the atmosphere. The beta decay and gamma decay will lose their energy long before any man will drink the water.

The greater concern with radiation and drinking water is the presence of radioactive isotopes in the water. If radioactive isotopes are in the water then these radioisotopes will continue to emit decay. The amount of time that the radioisotopes in the water will be a health problem will depend on 1) the half-life of each radioisotope, and 2) the initial amount of each radioisotope in the water.

Ingestion of Radiation Through Food

As with water, if radioactive decay does hit any food, then that decay will not last long. The more significant concerns are radioisotopes in the nearby water or soil. If radioisotopes are in the water or in the soil, then nearby agriculture might absorb those radioisotopes. The agriculture will then incorporate the radioactive isotopes throughout the roots and stalks (much as the process of humans inhaling radioisotopes and spreading those via the blood).

Penetration of Each Type of Radiation

There is a difference between exposure to radiation and radiation actually entering the body. You may be exposed to radiation and yet not have radioactive decay enter your body. Whether or not radioactive decay enters your body depends on primarily on the size of the decay.

An alpha particle (Helium atom) is too large to enter your skin. At most, alpha decay will penetrate no further than a few micrometers into human skin.

Beta decay (ejected electron) is more penetrating than alpha. Electrons are small and so they can enter into the body more easily than the alpha particles. However, even with their small size, electrons tend to go only a limited distance in a human body (at most only a few millimeters).

Gamma decay is the most penetrating. Because gamma is small it can enter your body easily. Gamma radiation can easily travel through the skin, and often reach into the internal organs. Gamma decay is the main form of radiation that causes cancer.

Table #1: Human Penetration and Shielding Requirements

Name	Penetration in humans (max)	Shielding
Alpha particle	.05 mm	paper, clothing
Beta particle	4 mm	wood, plastics, thin metal
Gamma ray	deep penetration possible	lead, concrete

Energy of Each Type of Radioactive Decay

Introduction

The amount of damage caused by radioactive decay is directly related to the amount of energy of the decay. The greater the amount of energy, the more damage the decay can cause.

When discussing the effects of radioactive decay, we must consider two locations of energy measurements. These locations are: a) the energy when the decay is emitted (initial energy), and b) the energy when the decay hits a specific cell (energy before impact).

Taking these concepts together, we have four factors to consider:

A. <u>The initial energy of the decay depends on</u>:
    1. The type of decay
    2. The specific isotope

B. <u>The energy of the decay before impact depends on</u>:
    3. The loss of energy as the decay travels through air
    4. The loss of energy after the decay hits other molecules

1. Initial Energy: Type of Decay

The initial energy of radioactive decay can range quite a bit, depending on the type of particle and the particular isotope. Overall, the initial energy of radioactive decay is in the MeV (million electron volt) range. Remember that each MeV is equal to 1.6×10^{-13} Joules.

Table #2: Energy of Radioactive Particles (in MeV)

Name	Energy (possible range)	Energy (most common)
Alpha particle	2 MeV to 7 MeV	4 MeV to 5 MeV
Beta particle	.05 MeV to 4 MeV	.22 MeV to 1.8 MeV
Gamma burst	.1 MeV to 10 MeV	1 MeV

Table #3: Energy of Radioactive Particles (in Joules)

Name	Energy (possible range)	Energy (most common)
Alpha particle	3.2×10^{-13} J to 1.1×10^{-12} J	6.4×10^{-13} J to 8.0×10^{-13} J
Beta particle	8.0×10^{-15} J to 6.4×10^{-13} J	3.5×10^{-14} J to 2.8×10^{-13} J
Gamma burst	1.6×10^{-14} J to 1.6×10^{-12} J	1.6×10^{-13} J

2. Initial Energy: Specific Isotope

The initial energy of the ejected radioactive decay depends on the particular isotope. Note that the energy of the decay can vary even among isotopes of the same element. For example, both radioactive isotopes of Calcium emit beta particles (electrons), yet the electrons emitted have different energies:

Isotope	Decay	Energy
Calcium–45	beta	.25 MeV
Calcium–47	beta	1.9 MeV.

Therefore, the electron from Calcium-47 (1.9 MeV) can do more damage than the electron from Calcium-45 (.25 MeV).

3. Energy Before Impact: Loss of Energy as Decay Travels Through the Air
Every radioactive decay loses energy as it travels. This means that the further away you are from the source of decay, the safer you are.

Alpha – As the alpha particle hits molecules in the air, then that alpha particle loses energy. The alpha particle eventually slows down to a safe energy level. In dense air, alpha particles will travel only a few inches before becoming safe.

Beta – When a beta decay particle (electron) hits the air, the beta particle ionizes one of the air molecules. The beta particle loses some of its energy in the process. After a few of these encounters, the beta particle does not have enough energy to cause any more ionization. By the time the electron reaches you, it has so little energy that it is of no harm to you at all.

In the most ideal traveling circumstances (such as spacious, low density air) the electron will travel only a few feet before becoming safe. As for denser materials, electrons from beta decay tend to go only a few millimeters into a dense material before becoming non-energetic.

Gamma – Gamma decay is a pulse of electromagnetic energy. As with all electromagnetic waves, the energy spreads out as it travels. (You can see this phenomenon when you use a flashlight). Eventually, the gamma energy pulse spreads out so much that anyone who comes in contact with the gamma energy will not be affected.

4. Energy Before Impact: the Loss of Energy Each Time Decay Hits a Cell
As an alpha or beta particle goes bouncing around, that particle has less energy after each hit. With each successive hit, the particle imparts less energy, and therefore, causes progressively less damage. Eventually the particle will be completely safe.

However, gamma decay is different. Once gamma hits a cell, that energy is generally absorbed. However, in the case of gamma decay we can accurately calculate the energy as it relates to distance. This will be discussed in the section below.

Energy Spread of Gamma as Related to Distance

Introduction

The energy of each type of radioactive decay decreases as it travels further from the source. At a certain point, all decay becomes safe. For gamma decay, we can accurately calculate that distance. This method can be useful because the gamma radiation is the most harmful.

Spreading Electromagnetic Energy

Any burst of electromagnetic energy will spread out as it travels. We commonly see this effect in a flashlight and in the stars. When we hold a flashlight, we can see the light spread out. However we also see that the light is dimmer in the distance. Looking just a few feet in the distance, the flashlight provides no useful light at all.

We have a very similar effect with the spread of gamma radiation. Remember that gamma decay is a pulse of electromagnetic energy. Therefore, gamma energy spreads out as it travels. This is just the same as a beam of light spreading out and becoming dimmer as it travels.

Furthermore, just as there is a distance at which an observer can no longer see a beam of light, eventually the pulse of gamma energy spreads out so much that the energy at any one location is very small.

Energy Spread and Impact on Humans

If you stand next to the sun, then you will get so much electromagnetic radiation that you will burn. Yet, if you stand millions of miles away, as we do on Earth, then the electromagnetic radiation is pleasant. Similarly, if you stand next to the initial burst of gamma radiation from an isotope, then you may be seriously harmed. However, if you are miles away from the initial gamma burst, then you will get so little energy that your body will not be affected.

The Inverse Square Law

The amount of energy from a gamma burst at any location can be calculated using the "inverse square law." The inverse square law applies to any wavelength of electromagnetic energy. Using this rule, we can calculate the fraction of the original energy amount (and hence the final energy value), at any location, as related to the distance traveled.

The law is written as: Energy fraction = $1/(distance)^2$. In order to use this tool, we must first pick a distance from the source, then we must measure the energy at that location. After that is done, we can then pick any other distance and calculate the energy at that location. (There is no need to measure anymore).

Note that the first distance we pick must necessarily be further than the length of the electro-magnetic wave. For visible light and gamma rays, the wavelengths are so short that this is not a problem. A distance of 1 meter will be convenient.

When using the inverse square law, energy is calculated in terms of fractions. For example, twice the original distance would result in 1/4 the original energy:

  • Inverse Square law: Energy fraction = $1/(distance)^2$
  • We put in a "2" for the distance, since we want to know the final energy for twice the original distance.
  • Therefore, the energy fraction = $1/2^2 = 1/4$ of the original energy

Example of Energy Spread for Gamma Burst

Let us assume that we measure the energy of gamma decay at exactly 1.0 meters from the source. Let us also assume that the measured value of the gamma burst (at a distance of 1.0 meters from the source) is 6.0 MeV. Therefore, using the inverse square law, we would get the following energy spread values:

Distance	Energy fraction	Final Energy
1 meter	1	6.0 MeV
2 meters	1/4	1.5 MeV
3 meters	1/9	.66 MeV
4 meters	1/16	.37 MeV
5 meters	1/25	.24 MeV
and so on...		

Remember these concepts: the further that a gamma burst travels, the more it spreads. The more the energy spreads, the safer it becomes. This also means that the further away a person is from the source, the safer he will be.

Chapter Summary

1. There are four routes of entry for radiation to enter your body. However, not all of these routes are equally likely to occur.
   a. Traveling through the skin
   b. Inhalation
   c. Ingestion in the drinking water
   d. Ingestion in the food

2. Whether or not radioactive decay enters your body depends primarily on the size of the decay. Smaller particles are more likely to enter your body than large particles.

3. An alpha particle (Helium atom) usually gets no further than a few micrometers into human skin.

4. Beta decay (ejected electron) is more penetrating than alpha. However, even with their small size, electrons tend to go only a few millimeters in a human body.

5. Gamma decay is the most penetrating. Because gamma is small, it can easily travel through the skin, and often reach into the internal organs.

6. All three types of radioactive decay ionize the cells. However, each type of radioactive decay does this in different ways. Alpha and beta particles are like pinballs: these particles go bouncing around, hitting cells, and ionizing each molecule that the particle hits. Gamma radiation is like a knife. Gamma radiation is highly localized, but it is certain to cause significant damage.

7. When a cell becomes damaged several things may happen:

    a. The cell repairs itself completely.

    b. The cell stays permanently damaged, but does not affect the other healthy cells. The extent of the health problem depends on which cells were damaged, and how many.

    c. The DNA of the cell becomes damaged, which creates cancer.

8. Beta particles tend to cause skin damage commonly referred to as radiation burns.

9. The amount of damage caused by radioactive decay is directly related to the amount of energy of the decay. The greater the amount of energy, the more damage the decay can cause.

10. The initial energy of the decay depends on:

    a. The type of decay

    b. The specific isotope

11. The energy of the decay at a particular cell depends on:

    a. The loss of energy as the decay travels through air

    b. The loss of energy after the decay hits other molecules

12. Every radioactive decay loses energy as it travels. This means that the further away you are from the source of decay, the safer you are.

13. Alpha particles and beta particles lose energy by colliding with other molecules.

14. Gamma decay loses energy because any burst of electromagnetic energy will spread out as it travels.

15. In the case of gamma decay we can accurately calculate the energy as it relates to distance. The amount of energy from a gamma burst at any location can be calculated using the inverse square law:

    Energy fraction = $1/(distance)^2$.

7.11
Measuring Radiation

Introduction

Measuring radiation should be simple. Unfortunately, measuring radiation is far more complicated than it should be. There are two fundamental problems with existing radiation measurements: 1) the units of radiation measurements are inadequate, and 2) the units of radiation measurements are complicated. These two fundamental problems lead to miscommunication when trying to compare amounts of radioactive decay.

In the following sections we will explain everything related to measuring radioactive decay. In particular, we will discuss the advantages and limitations of each method.

Physical vs. Biological Radiation Measurements

When we first consider radiation measurements, there are two broad categories: 1) physical properties and 2) biological effects.

Physical properties are qualities of the radiation itself. The main physical property for radiation measurements is energy. Other physical properties can be used such as type of decay and size of decay. Physical properties are ideal for measuring the radiation in the air. However, at this time physical properties are not often used.

Biological effects are the results of radioactive decay on the human body. The most common units of radiation measurements are those units which measure biological effects. In fact, any term you read regarding radiation measurement is likely to be a measurement of biological effects (rather than a measurement of a physical property). Biological effects are usually measured in terms of energy absorbed per mass of living tissue. For a biological unit to be used, the radiation must be absorbed by the body. Note that biological units are useless when measuring radiation in the air, soil, or water.

Physical properties are more universal than biological properties. This is because physical properties are qualities of the radiation itself. In contrast, biological effects are what the radiation does to an organism. Biological effects can only be measured when the decay is absorbed by man.

Physical Radiation Measurements

Introduction

Physical properties are qualities of the radiation itself. The main physical properties for radiation measurements are type of decay and energy of decay. As stated above, measuring radioactive decay based on physical factors is preferred over biological measurements because physical factors are more absolute. Physical factors are also much more accurate when measuring the amount of radiation in the air.

Proposed Physical Radiation Measurements

Radiation measurements based on physical properties are important, for it is through these that we can accurately measure radiation in the air, soil and water. However, at this time physical measurements are rarely used. Biological measurements are used far more frequently.

There are nuclear engineers who have observed this inadequacy. Some engineers have proposed their own ideas for measurements. However, these ideas have yet to become popular among the scientific community. The most common proposed physical measurements are as follows:

    a. Total energy, per volume of space (Joules/m^3), for alpha decay

    b. Total energy, per volume of space (Joules/m^3), for beta decay

    c. Total energy, per volume of space (Joules/m^3), for gamma decay

Physical Radiation Measurements: Types of Decay

When measuring radioactivity, it is important to measure each type of radioactive decay. There are two important practical issues based on the different types of decay to consider: 1) different types of decay are emitted depending on the specific radioisotopes in the mix, and 2) each type of radioactive decay has different penetration abilities (due to the different sizes of each decay). Therefore, measurements of radiation should be specific to each type. However, at this time, radiation measurements do not usually specify decay type.

Note that this requires three different detectors. Each type of decay is a different physical object. Therefore we need three different devices to detect the presence of each one. (The three detectors can be devised in the same box, but it is important to remember that three different detection mechanisms are needed).

Physical Radiation Measurements: Energy

Ultimately, the best overall factor of radiation to measure is energy. Remember that greater energy will result in more damage. Therefore it is essential that we measure the energy of radioactive decay.

There are many decisions to make regarding measuring the energy of radioactive decay. Scientists debate the specifics of exactly how this should be done. The broad questions are:

1. Which types of decay should we measure?
2. Where should we measure the decay?
3. How often should we measure the decay?
4. What tools should we use to measure the decay?

Measuring Decay Energy: Which Types of Decay?

Which types of decay should we measure for energy? Remember that three different devices are needed to measure the three different types of decay. Also remember that the device must not only measure for type of decay, but for energy of decay. We also will want to place devices in several different areas. Therefore cost may be a factor when choosing devices. We may need to be selective.

We should measure gamma, because gamma is the most penetrating and causes the most damage. We might choose to measure beta, because beta can cause skin lesions. However, we might not need to measure alpha, because alpha will not enter the skin.

Measuring Decay Energy: Where to Measure?

Where should we measure the energy of each decay? The answers to that question are debated among scientists. However, here are some guidelines (as collected from various sources) regarding measurement locations:

- We should not just measure the energy at one location. This value would give a false impression to those who live at distances which differ from that location.

- We want multiple locations for accuracy, but we do not want measurements to become too cumbersome or too expensive.

- Some of the proposed locations of energy measurements include: inside the plant; immediately outside the plant; and 5 miles from the plant.

- Energy measurements should always include three specific values, one energy value for each type of decay.

Measuring Decay Energy: How Often?

How often should we measure the decay? The answer depends on the location and the purpose. The measuring devices on the property of the nuclear power plant should measure the energy of radioactive decay continuously. The information should be analyzed hourly.

Local government offices should measure radiation during peak hours of power plant production. They offices can limit their analysis to daily review.

Any other entity (news organizations, universities) can measure and analyze as needed and as their budget permits.

Measuring Decay Energy: What Devices?

What equipment should be used to measure the energy? The answer to this question is outside the scope of this book.

Note that we will discuss the most common devices for measuring radiation (Geiger Counter and Scintillator) later in this book. However, these devices do not measure energy.

Biological Effect Measurements

Introduction

The main reason we care about radiation is for the health of humans. Therefore, many people choose to measure radiation in terms of the final result: biological effects. However, there are problems with measuring radiation for biological effects. The two main problems are 1) no common language, and 2) the common practice of combining all types of decay into one measurement.

One of the major problems with the measurements of biological effects of radiation is that there are many different units. With so many units, it can be difficult to compare data.

The most common units are described below. Some of these units will require further explanation, particularly those which use "equivalents." Those details will be provided in the next section.

The other major problem with the measurement of biological effects is the practice of combining all decay into one value. There are three types of decay. Each type has different penetrating ability and each type has different energy. Therefore the three types of decay should be treated separately. Yet it is common practice among those who study biological effects of radiation to combine all three types of radioactive decay.

The reasoning for combining the types of decay is that all three types of radiation affect people in generally the same way: by ionizing cells. However, some scientists believe that this practice of combining types of decay into one final value is a mistake. Although all types of radiation ionize cells, the types of decay are too different to make accurate comparisons on biological effects.

roentgen (R)

Definition: The roentgen is the amount of radioactive decay which creates $1.8 \times 10^{+12}$ ion pairs per gram of tissue. The roentgen was originally devised as a measurement unit for use with X-rays and gamma rays.

Alternate Definition: In order to make the roentgen sensible for radiation in air, some scientists have recently created an alternate definition for the roentgen. For radiation in air, the roentgen is the amount of radioactive decay which creates $2.0 \times 10^{+9}$ ion pairs per 1 cm^3 of dry air.

gray (Gy)

Definition: One gray corresponds to the transfer of one joule of energy to one kilogram of living tissue: 1 gray = 1 Joule absorbed/kg tissue. Note that the gray is used as the basis for the sievert. (See "sievert" below).

rad

   Definition: Like the gray, the rad is a unit of how much energy has been absorbed by living tissue. The term "rad" is an abbreviation for radiation absorbed dose. One rad is exactly equal to 0.01 gray. Therefore, 1 rad is exactly .01 joules of energy absorbed by a kilogram of tissue: 1 rad = .01 Joule absorbed/kg tissue. Note that for gamma rays and X-rays, 1 rad = 1 roentgen. Also note the rad is used as the basis for the rem. (See "rem" below.)

rem

   The rem is a "dose equivalent." (Equivalents will be discussed fully in the next section.) The unit "rem" is an abbreviation for roentgen equivalent in man. Some sources state that the abbreviation is for radiation equivalent in man. Many people today prefer the unit of "rem." However, the Sievert (below) is also a preferred unit by many people.

Definition 1 (rem as related to the rad)

   The rem is the dose of any type of radiation (alpha, beta, or gamma) which has the same health effect (equivalent biological effect) as one rad of gamma radiation. The rem starts with the unit of the rad, which is .01 joules of energy absorbed by a kilogram of tissue. This rad value is multiplied by certain factors (depending on the situation) in order to make an "equivalent."

   1 rem = 1 rad x equivalent dose factors
   = [.01 Joule/kg tissue] x [equivalent dose factors]

Definition 2 (rem related to the roentgen)

   The most technical definition of rem is usually stated in reference to the roentgen. Specifically, the rem is any type of radiation which in man has the same health effect as one roentgen of X-ray or gamma radiation.

Sievert, Sv

Definition: Like the "rem", the "sievert" is a dose equivalent. (The concept of "equivalents" will be discussed in the later.) Briefly, the sievert is the dose of any type of radiation (alpha, beta, or gamma) which has the same health effect (equivalent effect) as one gray of gamma radiation.

1 sievert = 1 gray x equivalent dose factors

= [1 Joule/kg tissue] x [equivalent dose factors]

The primary difference between the sievert and the rem is that the sievert starts with the gray (and then makes an "equivalent") whereas the rem starts with the rad (and then makes an "equivalent").

Further note that because 1 gray = 100 rad, then the sievert (which is based on the gray) can be related to the rem (which is based on the rad) as: 1 sievert = 100 rem.

Biological Effect Measurements: Equivalents

Introduction

Earlier we said that radiation measurements can be complicated because of the concept of "dose equivalent." In brief, the concept of "equivalents" takes the number of alpha and beta decay, and turns that into an "equivalent" number of gamma decay. The logic is faulty, the comparisons are guesswork, yet the practice is very common.

Absorbed Dose and Dose Equivalent

The "absorbed dose" is the straight-forward value of energy absorbed per mass of tissue. The gray and the rad are units of absorbed dose. In contrast, the "dose equivalent" is the equivalent biological effect as if the only decay were gamma. The sievert and the rem are units of "dose equivalent." These units require some conversion factors.

The Equivalence Calculation in General Terms

The general dose equivalence calculation starts with the unit of the gray or the rad. The starting value for the calculation is the number of joules of energy absorbed by the body, per kilogram of tissue. This value of energy per body mass is then multiplied by a series of factors.

The net result, ideally, is designed to arrive at an equivalent energy value in terms of gamma decay (as if all the absorbed radiation were gamma).

There are often other conversion factors besides type of decay. Depending on what biological effects are being studied, any one of several "equivalent" factors may be inserted into the conversion calculation (depending on what is being compared). The most common equivalent factors include:

1. decay type; and making equivalent to gamma.
2. species of animal (usually making equivalent to man)
3. part of the body absorbing the radiation (noting specific organs)
4. time exposed to radiation source

Logic of Equivalence

Logic must be applied when thinking about equivalence. Dose equivalent units, such as the rem, are trying to compare alpha particles and beta particles to gamma waves. This would be like trying to compare the effect of eating apples and oranges by comparing the amount you ate to some number of bananas. Such comparisons are difficult to do. Furthermore, the conversions are based on many factors, and therefore getting any precision is difficult.

There are many scientists who believe that comparing alpha decay to gamma decay is not a valid comparison. Similarly, many scientists also believe that comparing beta decay to gamma decay is not a valid comparison.

In addition, alpha decay is not likely to enter the body, and beta decay can be prevented by using simple protection. Thus, you might be exposed to a lot of alpha decay or beta decay, but because you might not have absorbed any alpha or beta you can have a high exposure and yet not have been affected.

Dosages and Biological Effects

Introduction

There are many studies of radiation absorption and health effects. The data varies from report to report. This variation in data is not so much a variation in biology as it is a variation in how the measurements were taken and what units were used in the final report.

However, to get an approximate understanding of radiation dosage and biological effects, the data table below offers a composite of several reports with values reported in the units of rem. As you read the data below, keep in mind the following:

1. In order to have a sense of the "rem", know that if the rem is only gamma radiation, then 1 rem = .01 Joules energy absorbed per kilogram body.

2. Also note that the "dosage" value assumes a) the energy of radiation was absorbed by the body, and b) this total amount of radiation was absorbed at one time.

Approximate rem value (dose equivalent) and biological effect

• 0 to 25 rem absorbed: No detectable clinical effect in humans.

• 25 to 100 rem absorbed: Slight reduction in number of blood cells, short-term.

• 100 to 200 rem absorbed: Longer-term reduction in number of blood cells.

• 200 to 300 rem absorbed:

 Nausea first day of exposure; sickness after two weeks; Recovery in about three months unless complicated by infection.

• 300 to 600 rem absorbed:

 Nausea in first few hours, sickness in 1-2 weeks. Some deaths in two to six weeks. Eventual death for 50% if exposure is above 450 rem; others recover in about six months.

• 600 rem absorbed and greater: Sickness in first few days, eventual death of nearly 100%.

Unit Comparisons of Biological Effects

In this section we will use the same dose-effect information above, but put the data in context with other units.

Table #1: Dose ranges in terms of rem, Sv; and effects in brief

Dose range	Rem	Sievert	effects (brief)
range #1	0 – 25	0 – 2,500	no effect
range #2	25 – 100	2,500 – 10,000	blood cells down, short term
range #3	100 – 200	10,000 – 20,000	blood cells down, long term
range #4	200 – 300	20,000 – 30,000	sickness, recovery in 3 months
range #5	300 – 600	30,000 – 60,000	some recovery, some deaths
range #6	above 600	60,000 and above	death almost certain

Locating Radiation: The Geiger Counter

Introduction

The Geiger counter is used to find Uranium deposits in the earth, as to test for leaks in storage containers. Another common tool for this purpose is the scintillator, which is similar to the Geiger counter.

However, the Geiger counter provides limited information. The Geiger counter can tell us that radiation exists, but it cannot tell us the type of decay. The Geiger counter can tell us where radiation is stronger, but it cannot tell us exact amounts of energy.

The essential concept of the Geiger counter is that radioactive decay hits a gas molecule, which causes the gas molecule become ionized. The ionized gas creates bursts of electrons in the wire, which can be conveyed as a series of clicks.

Basic Process of the Geiger Counter

The basic process of the Geiger counter is as follows:
1. The Geiger counter is a tube, with a wire in the middle, and filled with gas.
2. Radioactive decay passes through the container and into the tube.
3. When radioactive decay hits a gas molecule, the gas molecule becomes ionized. The respective ions have their effects on the wire, causing a burst of current to flow.
4. The final current is converted into a sound, which we hear as a "click."
5. Many Geiger counters also have a display. The readout is usually given in terms of "clicks per minute."

Multiplying Ions in the Geiger Counter

One ion pair would not be enough to make a signal. However, the ion pairs naturally multiply. The gas ions have enough energy to ionize other gas molecules. By the time a gas ion reaches the electronics and causes an electron to flow, there are dozens of gas ions doing the same thing. (This is known as the "avalanche effect.") The net result is a signal that can be registered from the impact of a single decay.

Types of Decay which Trigger the Counter

There are three types of decay, yet each affects the Geiger counter differently. The design (and expense) of the specific Geiger counter will determine how well each type of radioactive decay will trigger the counter. All Geiger counters will be triggered by beta decay (electrons). Most Geiger counters will be triggered by both gamma decay and beta decay.

Geiger counters usually are not triggered by alpha decay. Alpha decay (helium nucleus) is usually absorbed by the outer container of the Geiger counter. Therefore, most Geiger counters will not register alpha decay. However, some of the more expensive models are made of materials which allow alpha decay to pass through, and therefore trigger the counter.

Usefulness of the Geiger Counter

The Geiger counter has two primary uses:

1. Finding locations of radioactive decay. For example:
    • Radioactive decay is in room 6, but not in room 7.
    • Radioactive decay is leaking through storage container #23.

2. Comparing relative amounts of decay. For example:
    • the radioactivity has decreased since last week
    • the radioactive decay at this spot is much higher than normal

Limits of the Counter

The Geiger counter has several limits:

    a. The counter cannot distinguish between types of decay

    b. The counter cannot count decay accurately

    c. The counter cannot measure energy

It is important to note that the Geiger counter is able to count only the final current. The Geiger counter cannot distinguish between types of decay. Nor can the Geiger counter measure the amounts of each decay.

Remember that the burst of current in the counter is created essentially by gas ions. Although the gas is ionized by the radioactive decay, it is not the radioactive decay which creates the current. Therefore, the counter cannot identify decay type.

It is important to note that the Geiger counter cannot count decay accurately. The current is created not just by one gas ion, but by dozens (remember the "avalanche effect" discussed above.) We can never be certain how many gas ions were created per initial ionization. Therefore we cannot correlate number of clicks with a specific number of decays entering the tube. We can get a qualitative understanding, but we cannot get any accurate measurement.

It is also important to note that the counter cannot measure energy. All that is required for the Geiger counter to work is that the energy be high enough to ionize the gas. Any energy beyond that value cannot be determined.

The Scintillator

Another common tool for detecting radioactive decay is the scintillator, which is essentially just like the Geiger counter. (In fact, the scintillator is also called a scintillation counter.)

The scintillator differs from the Geiger counter in that the scintillator conveys radioactive decay as a pulse of visible light, whereas the Geiger counter conveys decay as a clicking sound. A secondary difference is that some scientists have found the scintillator to be more sensitive to gamma than the Geiger counter, and hence more effective to counting gamma decay.

The limits of the Geiger counter apply equally to the scintillator. This means that the scintillator cannot identify decay type, and the scintillator cannot measure energy of the decay.

The Film Badge

The film badge is a convenient method for a worker to know his approximate exposure to radioactive decay. The badge is indeed a film, usually made with silver halide, and works very similarly as camera film. In camera film, when the film is exposed to visible light the structure of the molecules will change. Similarly, in the film badge when the film badge is exposed to gamma waves the chemicals in the film badge will change.

In camera film you will get a better picture with exposure to more light, yet too much light will overexpose the film. In a similar fashion the film badge will respond to gamma waves. Greater exposure to gamma waves will produce a more noticeable result.

The changing color in the film badge is directly proportional to the amount of gamma waves which hit the badge. The net result is that the changing color of the badge will give the worker an approximate idea of how much gamma decay he has been exposed to.

Tracking Radioisotopes in the Environment

In many cases, measuring the decay is not enough. Sometimes we must also measure radioisotopes in the environment. Whether the radioactive isotopes exist naturally or exist due to nuclear power leaks, these radioactive isotopes should be located and quantified.

The first step must be to locate the decay. This is usually done with a Geiger counter. Sometimes it is necessary to identify the specific radioisotopes. This requires a special analytical tool, such as a mass spectrometer. After the radioactive isotopes have been located and identified, then other tools can be used to measure types of decay and their the energy levels.

In total, the steps required for measuring radioisotopes include:
1. Locate radioactive decay using a Geiger counter
2. Measure the total energy of the decay
3. Measure the energy of decay, by types of decay
4. Identify the types of decay
5. Quantify the types of decay
6. Identify and quantify specific radioisotopes

Other Radiation Measurements

Introduction

There are a few other units to note. You might hear or read these units, especially in older reports. These units are discussed at the end of the book because they are older units. Each of these units measure decay as an amount of decay per time, which is different from measuring physical factors or biological effects.

curie (Ci)

The curie is a unit of decay per time. The curie is defined as: 1 curie = 3.7×10^{10} decays per second. The curie is a count of decay per time. As with the Geiger counter, the curie does not specify type of decay, only that decay occurs.

becquerel (Bq)

The becquerel is a unit of decay per time. The becquerel is defined as: 1 becquerel = 1 decay per second.

decay per minute (dpm)

Exactly as it sounds, the dpm is rate of decay. As with the curie and the becquerel, the dpm does not specify type of decay.

Summary

1. There are two general categories of radiation measurements: physical properties and biological effects.

2. Physical properties for radiation measurement are more universal than biological effects. Physical properties are ideal for measuring radiation in the air. Physical properties include: type of decay and energy of decay.

3. The most common proposed measurements for physical properties of radiation are:
   a. Total energy, per volume of space (Joules/m^3) – alpha decay
   b. Total energy, per volume of space (Joules/m^3) – beta decay
   c. Total energy, per volume of space (Joules/m^3) – gamma decay

4. Measuring biological effects of absorbing radioactive decay is very common. However, these measurements have meaning only when decay is actually absorbed by the body. Such measurements are useless when measuring radiation in the air.

5. There are problems with the current methods of measuring radiation for biological effects. The two main problems with current measurements are the lack of common language, and the common practice of combining decay.

6. The most common units for measuring biological effects of radiation are: roentgen, gray, rad, rem, and sievert.

7. Units for radiation, measured by biological effects, include:
   - 1 roentgen = 1.8 x10^{+12} ion pairs created per gram of tissue
   - 1 gray = 1 Joule absorbed/kg tissue
   - 1 rad = .01 Joule absorbed/kg tissue
   - 1 sievert = 1 gray x equivalent dose factors
   - 1 rem = 1 rad x equivalent dose factors

8. Conversions for radiation units, measured by biological effects:
- 1 gray = 100 rad
- 1 rad = .01 gray
- 1 sievert = 100 rem
- 1 rem = .01 sievert

9. The absorbed dose is the straight-forward value of energy absorbed per mass of tissue. The gray and the rad are units of absorbed dose.

10. The dose equivalent is the equivalent biological effect as if the only decay were gamma. The sievert and the rem are units of dose equivalent. These units require conversion factors.

11. The dose equivalent calculation starts with the unit of the gray or the rad. This value of energy per body mass is then multiplied by a series of factors. The net result is designed to arrive at an equivalent energy value in terms of gamma decay.

12. There are many studies of radiation absorption and biological effects. However, we must compare the units carefully. We must also consider how the equivalent factors were calculated.

13. It is often necessary to locate radiation. Locating radioactive decay is best done with the Geiger counter.

14. The Geiger counter is useful for locating radiation and comparing relative amounts of decay.

15. The Geiger counter is limited. The Geiger counter can tell us that radiation exists, but it cannot tell us the type of decay. The Geiger counter can tell us where radiation is stronger, but it cannot tell us exact amounts of energy. The Geiger counter only be used for qualitative assessments, it cannot be used for any quantitative measurements.

16. Sometimes it is necessary to track specific radioisotopes. In order to track radioisotopes, the general location must be found such as by using a Geiger counter. Then a mass spectrometer can be used to identify specific radioisotopes in a sample.

7.12
Storing Nuclear Waste

Introduction

One of the concerns regarding nuclear power is the safe storage of nuclear waste. There are two storage concerns: hazards from the inside of the containers, and hazards from the outside of the container.

The hazards from inside the container arise from the decaying radioactive waste. The first priority is to contain the isotopes so that no radioactive isotopes escape into the air or water.

Secondly, as the isotopes decay they produce energy. This energy must be absorbed by the container effectively. Over a period of millions of years, the energy absorbed by the container may cause the container to crack, which would eventually allow the remaining radioactive isotopes to escape.

Note that there is one concern of hazards from within which cannot happen: radioactive explosion. It is physically impossible for a container of radioactive waste to explode.

Hazards from the outside are from weathering and terrorism. We often see the instant and total damage on various objects caused by tornadoes, hurricanes, floods, and earthquakes. Any one of these events would destroy a container of radioactive waste.

We have also seen the slow but powerful effect of erosion, such as caused by running water, high winds, or shifting soils. As we consider storing containers for millions of years, we must remember that such long-term erosion can cause cracks in the containers.

Finally, we know that terrorism is a very real threat. Many containers are currently stored above ground; just one small bomb dropped from a plane would destroy a dozen containers, releasing radioactive waste into the air.

Clearly there is no doubt that nuclear waste must be stored for a long time. The question then becomes what is the best method for storage?

Steps in Storage

Typical storage of nuclear waste includes the following steps:

1. The used nuclear fuel pellets and the by-products of fission are kept in a large pool of water. These isotopes are kept in this pool for water for at least a year. This is done so that the nuclear waste can cool completely.

2. The isotopes are vitrified into glass-like beads. The radioactive isotopes are essentially in a solid form. This makes it more difficult for the radioactive decay to seep into water or escape into the air.

3. The vitrified by-products are then sealed in heavy containers, usually steel or ceramic. Steel containers are most common. The material of steel prevents all decay (including gamma which is the most penetrating decay) from escaping the container.
   Note that steel can corrode. Therefore, the outer layer of the steel is a special alloy or a ceramic which is resistant to water. The advantage of ceramics is that ceramics do not corrode.
   Some ceramics can hold radioactivity as effectively as steel. Sometimes there is a steel container (as described above) which is completely encased in a ceramic container.

4. These containers usually sit above ground, as temporary storage.
   Until such time as an underground facility is built, these containers sit above ground, on a site nearby the power plant.

5. The containers are eventually buried deep into the ground.
   The containers will eventually need to be buried, due to weathering. Sites are chosen which: have no water in the ground, are geologically very stable, and are far away from any community.

Long Term Storage: Inside Dangers

Introduction (the hazards)

The hazard of nuclear waste depends on what is being stored. Inside each container is a mixture of radioactive isotopes, and therefore inside each container is a mixture of potential hazards. Each radioactive isotope has its own decay time. Many isotopes will decay within a few years. However, some isotopes will require thousands or millions of years to decay. Therefore, the storage system must contain the nuclear waste for a long time.

Heat Generated by Decay

Nuclear waste can raise the temperature of the containers. It is important to understand how this rise in temperature occurs so that we can design the storage system to handle this heat.

Every decay will have some energy when emitted. When the decay hits the walls of the container, the containers will absorb that energy. Recall that temperature is simply a measure of kinetic energy. Therefore, as the container absorbs the energy of the decay, the container will become progressively hotter. How hot will these containers become? This depends on many factors, including: overall initial energy of the decay, heat capacity of the container, thickness of the container, and cooling mechanisms. A few notes on each factor will be stated below.

1. Overall Initial Energy of the Decay

A greater initial energy of the decay, will result in a higher the temperature of the inside walls of the container.

2. Heat Capacity of the Container

Heat capacity is the ability of a material to absorb energy without raising the temperature. Metal cannot hold much energy before the temperature goes up. Therefore the steel walls of the container will rise in temperature very quickly. In contrast, ceramics can hold a significant amount of energy before the temperature goes up. Therefore, the ceramic containers will absorb the energy for some time without any significant rise in temperature.

3. Thickness of the Container

With a thicker the container, more energy can be absorbed by the container before the temperature will rise. More important, the thicker the container, the more energy that can be absorbed by the container before the outside becomes hot.

4. Cooling Mechanism

If the containers can be cooled from the outside (or if the hot air can be taken away), then the container walls will last longer.

The technology for containing the inside dangers of nuclear waste is well developed at this time. The nuclear waste will not leak through the container by itself. (And it is physically impossible for the nuclear waste to explode). Therefore, if there were no damage to the container from the outside then these containers would be effective in containing nuclear waste safely, over a period of millions of years.

Long Term Storage: Outside Dangers (weathering)

Introduction

The main hazard to nuclear waste containers comes from the outside. Weathering can damage any material, given enough time. Therefore, the nuclear waste containers will need to be protected from the forces of the natural world.

There are several types of outside dangers. The most significant of these are: heat, rain, snow, hail, tornadoes, earthquakes, landslides, fires, and terrorism. Heat from the sun and from fires will cause containers to crack. Water will erode the ceramic containers and rust steel containers. Hail will dent the containers. Earthquakes, even minor quakes, will cause cracks in the containers. Landslides will destroy containers by the weight of the earth. Tornadoes will rip containers apart. Acts of terrorism will destroy containers if the containers are hit directly.

Most nuclear waste is safe, for the moment. However, even under the best conditions the outside dangers will eventually destroy any container. The primary question is how long will these containers be safe if they are exposed to the weather?

<u>Outside Dangers: The Solutions</u>

In order to prevent the containers from eroding over thousands of years, we must put these containers in a location where nothing happens. This means that we must satisfy the following requirements:

1. This location must not be exposed to the outside. This will prevent damage from rain, heat, snow, hail, tornadoes, and terrorism.

2. This location must be geologically stable. This will prevent damage from earthquakes and landslides.

3. This location must be remote. This will be an extra precaution in case the containers ever become damaged enough to leak.

4. The location must be secure. This will prevent terrorist activities.

Note that some power companies have built bunkers to store their nuclear waste. However, the majority of waste containers sit above ground. This will be the situation until the underground site at Yucca Mountain is fully developed.

Yucca Mountain: Overview and Location

<u>Introduction</u>

Yucca mountain of Nevada will soon become a federal depository of nuclear waste. Yucca Mountain is one of the most well studied mountains in the world. Because this site has been considered as a repository, scientists have studied it thoroughly. These studies have spanned more than 20 years, and cover numerous topics.

There are several major reasons why the Yucca Mountain site is a wise choice for storing radioactive waste. Major reasons include:
1. Remote Location
2. Secure Area
3. Dry Climate
4. Won't affect water supply
5. Geologically stable

Remote Location

Yucca Mountain is located in a remote area of Nevada. This is an isolated location, entirely within an already existing Air Force base. No homes or businesses will be built nearby. The site is in Nye County. The site is approximately 90 miles northwest of Las Vegas, and 15 miles from the nearest business. This is a desert area, very near Death Valley. Except for the military base, no one lives in the area. No homes or businesses will be built nearby.

Secure Area

Yucca Mountain lies not only in a desert, but within the boundaries of a Federal Air Force Range. Yucca Mountain exists entirely within the Nellis Air Force Bombing and Gunnery Range. Furthermore, within that Air Force Range, and directly adjacent to Yucca Mountain, is the Nevada Test Site. The Nevada Test Site is where most testing of our nuclear bombs took place. (In other words, nuclear activity has already taken place here.)

Dry Climate

The climate is very dry. Yucca Mountain receives only 7 inches of rain per year. No rain means that the mountain will be stable (no erosion, no landslides). This also means that the containers will be protected from water (therefore no corrosion of the containers). Yucca Mountain also has an "unsaturated zone" above the tunnels. In an unsaturated zone, water can be absorbed. Water also travels slowly through these layers. This means that very little water will enter the repository.

Won't Affect Water Supply

The actual location of the stored material will be at least 1,000 feet above the water table. Therefore, it is very unlikely that nuclear material will enter into that water. Most importantly, the water table in Yucca Mountain does not feed into any major river. Some of the water may go to Death Valley, but that is the only location. None of the water used by citizens of Nevada ever comes from Yucca Mountain.

<u>Geologically Stable</u>

Yucca Mountain has been found to be geologically stable. There will be no earthquakes in this area. The mountain is not likely to shift.

Yucca Mountain: The Repository

<u>Introduction</u>

The natural environment will provide most of the protection for long-term storage. However, the repository will be designed with several features which add extra protection. The nuclear waste will be buried approximately 1,000 feet underground. (This is 1,000 feet of rock.) This distance is a far greater than radioactive decay could ever travel. The tunnels will also be reinforced, so there will be no chance of the tunnels collapsing on the containers.

<u>Tunnels and Protection from Water</u>

The tunnels will be carefully designed to prevent water from reaching the nuclear waste.

1. The tunnels are located beneath 1,000 feet of rock. Water will take a long time to travel to the tunnels.

2. Tunnels will be built away from fractures in the rock, because water will travel in these small fractures.

3. The containers will be protected by a drip shield (see below).

4. Tunnels will be built with a drainage system, so that any water that does enter the tunnel will be drain away from the nuclear waste.

<u>Containers and Drip Shields</u>

The nuclear waste will also be placed in a series of several containers. Should one container become damaged, in the next thousand years or so, then other containers will serve as protection. The first container is made of stainless steel. This container is placed inside a second container, made of a high quality corrosion-resistant material. A third protection is then built around these containers. This barrier is called the "drip shield." The drip shield protects the containers from water and falling debris.

Development of the Yucca site: a Timeline

In the late 1970s, several sites were discussed in the scientific community as possible locations for long-term storage of nuclear waste. Scientific studies began on each of them. In 1983 the Department of Energy selected nine sites to study in detail. In 1986 the DOE narrowed the list down to five sites, and they recommended three of these sites to the President. Scientific studies continued in earnest on those three sites.

In 2002, the DOE recommended Yucca as the best site for long-term storage of nuclear waste. This recommendation was given to the President, who submitted the recommendation to Congress. Congress then approved this site as a federal repository.

However, the site is not official yet. Although the site was approved by Congress, and although the site lies on Federal property, this site must still be approved by the Nuclear Regulatory Commission. At the time of publication of this book, the Yucca site is still waiting for official approval from the NRC.

The Federal Government is very open about explaining why the Yucca Mountain site is sensible. You can look at the websites, request brochures, request videos, and visit the information centers. For a limited time, you may also tour the mountain itself.

The final stage of the Yucca repository, that of driving nuclear material to the site at Yucca Mountain, was scheduled to begin in 2012. This date is being pushed further, to 2017 or 2020.

Driving the Nuclear Waste

Some citizens are concerned about the safety of radioactive material being driven from one location to another. The concerns are legitimate. However, the nuclear power industry has developed several safety precautions.

Whether driving fuel to the plant or waste away from the plant, several precautions are taken. For example, trucks will take special routes away from high traffic areas. Many trucks will also have GPS systems installed, which allows the truck's path to be tracked by safety engineers during the entire trip. Perhaps most important is the strength of the trucks. I have seen films of these trucks in tests (including crash tests, puncture tests, and nearby explosions) and I can assure you that these trucks are virtually indestructible.

Note also that the original nuclear fuel is driven *to* the nuclear power plants all the time and there has been no incident.

Dedicated trains, from power plants directly to the Yucca Mountain Site, are also being considered as a mode of transportation. Most of the railway would be entirely new tracks, and only used for hazardous cargo. However, the specific routes must be examined to evaluate the relative hazards.

In addition, sometimes these trucks and trains will be escorted by law enforcement and/or specially trained safety engineers.

Therefore, in general the transportation of nuclear waste is not likely to be a hazard, as long as the specially designed trucks are used, and both trucks and trains do not pass through populated areas.

The Future of Nuclear Waste

Using Radioactive Waste

In the future, we may find practical, safe uses for radioactive material. This has been the trend in other industries. For example, when oil is turned into gasoline there are by-products leftover. Some of these by-products were considered useless for many years. Today these by-products are used in products such as plastics.

Coal is another example. Ash is the biggest by-product of coal, and for many years the ash was just thrown away. Today ash is used in asphalt and concrete. Sulfur and Nitrogen, also by-products from coal power, were once discarded into the air. These chemicals are now collected and sold to chemical companies.

A similar result may happen for the nuclear power industry. We may, in time, find practical and safe uses for radioactive by-products. If we are able to do this, then we will have less radioactive waste to store.

Processing U-238 from Spent Fuel into Plutonium

One option for the future of nuclear waste is to separate the U-238 from the spent fuel and process it into Plutonium. The Uranium fuel pellets would be processed, without going to storage containers. Because the Uranium pellet is a high percentage of U-238, the U-238 is easy to locate. Separating U-238 requires a number of steps but U-238 can be isolated from the spent fuel pellets successfully.

This U-238 is then converted into Plutonium, which can then be used in nuclear reactors, as a second generation fuel. The remaining isotopes of the pellets would go to long-term storage.

There are several benefits of this proposal, including: more electricity is created per mass of Uranium, and less nuclear waste will need to be put into storage. The process has been proven to work, but the cost must be evaluated.

Other Options

There are other options for the future of nuclear waste being discussed. However, the practicality of these other options is still far in the future.

One option is to create new microbes which will consume the radioactive waste. However, the radioactive isotopes will still remain as they are (radioactive atoms) and therefore the microbes must be kept contained. In my view, this is more difficult than simply storing the vitrified waste in a container.

The ideal scenario is to separate useable fuels from the waste, thereby reducing the amount of waste and obtaining more energy from the mixture. However, separation methods are not yet practical.

Personally, I still believe that someday we will find practical, and safe, uses for the nuclear waste. I believe that someday other industries will use the waste from nuclear power, just as industries today use various waste from other processes.

Summary: Storage of Nuclear Waste

1. Typical storage of nuclear waste includes the following steps:
   a. Used nuclear fuel and the by-products are kept in a large pool of water, for at least a year, in order to cool.
   b. The isotopes are vitrified into glass-like beads.
   c. The vitrified by-products are then sealed in containers (made of steel, ceramic, or both)
   d. These containers sit above ground, as temporary storage.
   e. The containers will eventually be buried deep into the ground.

2. Nuclear waste must be stored for a long time. This is because of:
   a. Inside dangers (decay and energy of nuclear waste itself)
   b. Outside dangers (weathering of the container)

3. Nuclear waste must be stored for a long time because some of the radioactive isotopes continue to decay for many years.

4. Nuclear waste will not escape from the containers. This is because the waste is solid and the containers are made from steel or ceramics.

5. Nuclear waste cannot explode. This is physically impossible.

6. The natural world will weaken the containers (due to weathering). This may be instant, or take thousands of years, but it will occur eventually.

7. Due to the effects of the natural world, nuclear waste containers cannot remain above ground for more than a few years.

8. The Federal Government has been working on an underground repository for nuclear waste. Sites have been investigated for over 20 years. The final recommended site is Yucca Mountain.

9. The site at Yucca Mountain offers several advantages as a repository for nuclear waste, including: geologically stable, dry climate, won't affect water supply, remote location, and secure area.

10. Yucca Mountain is in the desert, and surrounded by a military base. The other side of Yucca leads to Death Valley. The area is dry, remote, stable, and secure.

11. The actual repository will be 1,000 feet below the ground, and 1,000 feet above the water table.

12. The repository will have a system of multiple containers and barriers.

13. Driving nuclear waste from power plants to the site at Yucca Mountain is a legitimate safety concern. However, many precautions will be taken.

14. The future of nuclear waste might include:
    a. finding a practical, safe use for radioactive material
    b. reprocessing U-238 from spent fuel into Plutonium

Conclusion

Many Americans hold passionate views about electrical power, yet few Americans understand all the details behind their passion. Electricity should not be mysterious. The science, the technology, and the data of electrical power can be understood by anyone.

Above all else, we must remember that there are no perfect solutions, there are only choices. Any option can be beneficial, yet each option has its own technical issues to work with. It is up to you and to your community to make those educated decisions. I hope that this book will help guide you in your choices.

M.F.

Appendices

A.5.1 Radioisotopes: decay type, what atom becomes, and half-life

A.5.2 Half-lives listed in order of decay time

A.5.3 Half-lives and number of steps

A.5.4 Multiple Decay Radioactive Isotopes

A.5.5 Decay Sequences for Multiple Decay Isotopes

A.5.6 Radioactive Decay Summaries

A.5.7 Suggested Nuclear Standards from ANS

A.5.8 Suggested Nuclear Standards from NRC

A.5.1   Radioactive isotopes, Decay type, and Half-lives

Points for Clarification:

1. This is a complete list of known radioactive isotopes. However this list does not necessarily mean that all of these isotopes are produced in a nuclear power plant. Which radioactive isotopes are created as a result of nuclear power? We don't know for certain at this time. Scientists must do more studies in order to answer that question.

2. All of these radioisotopes occur naturally. The only exceptions are the isotopes of Neptunium and Plutonium.

3. The existence of a radioisotope does not necessarily mean an abundance of that isotope. The isotope may exist only in trace amounts.

4. Isotopes are listed by atomic number, and then by isotope number.

5. The atomic number of each new element is in parentheses, next to the symbol for that new element.

6. The half-life can be anything from less than a second to billions of years. Read the units of time carefully.

Radioactive Isotope Data Table

Atomic #, Element	Isotope	Decay Type	Becomes	Half-Life
1 (Hydrogen, H)	H-3	beta	He(2)-3	12 years
4 (Beryllium, Be)	Be-10	beta	B(5)-10	1,600,000 years
6 (Carbon, C)	C-14	beta	N(7)-14	5,730 years
7 (Nitrogen, N)	N-13	beta	O(8)-13	10 minutes
8 (Oxygen, O)	O-15	beta	F(9)-15	2 minutes
9 (Fluorine, F)	F-18	beta	Ne(10)-18	2 hours
11(Sodium, Na)	Na-24	beta, gamma	Mg(12)-24	15 years
12 (Magnesium, Mg)	Mg-28	beta, gamma	Al(13)-28	21 hours
14 (Silicon, Si)	Si-32	beta	P(15)-32	100 years
15 (Phosphorus, P)	P-32	beta	S(16)-32	14 days
15 (Phosphorus, P)	P-33	beta	S(16)-33	25 days
16 (Sulfur, S)	S-35	beta	Cl(17)-35	8 days
17 (Chlorine, Cl)	Cl-36	beta	Ar(18)-36	310,000 years
18 (Argon, Ar)	Ar-39	beta	K(19)-39	269 years
19 (Potassium, K)	K-40	beta, gamma	Ca(20)-40	1,280,000,000 yrs
19 (Potassium, K)	K-42	beta, gamma	Ca(20)-42	12 hours
19 (Potassium, K)	K-43	beta, gamma	Ca(20)-43	22 hours
20 (Calcium, Ca)	Ca-45	beta	Sc(21)-45	5 1/2 months
20 (Calcium, Ca)	Ca-47	beta, gamma	Sc(21)-47	4.5 hours

21 (Scandium, Sc)	Sc-44	beta, gamma	Ti(22)-44	4 hours
21 (Scandium, Sc)	Sc-46	beta, gamma	Ti(22)-46	84 days
21 (Scandium, Sc)	Sc-47	beta, gamma	Ti(22)-47	3 days
25 (Manganese, Mn)	Mn-56	beta, gamma	Fe(26)-56	2.5 hours
26 (Iron, Fe)	Fe-59	beta, gamma	Co(27)-59	45 days
26 (Iron, Fe)	Fe-60	beta	Co(27)-60	100,000 years
27 (Cobalt, Co)	Co-60	beta, gamma	Ni(28)-60	5 years
28 (Nickel, Ni)	Ni-63	beta	Cu(29)-63	100 years
29 (Copper, Cu)	Cu-64	beta, gamma	Zn(30)-64	13 hours
29 (Copper, Cu)	Cu-67	beta, gamma	Zn(30)-67	3 days
31 (Gallium, Ga)	Ga-72	beta, gamma	Ge(32)-72	14 hours
32 (Germanium, Ge)	Ge-77	beta, gamma	As(33)-77	11 hours
33 (Arsenic, As)	As-74	beta, gamma	Se(34)-74	18 days
33 (Arsenic, As)	As-76	beta, gamma	Se(34)-76	26 hours
34 (Selenium, Se)	Se-75	beta, gamma	Br(35)-75	4 months
35 (Bromine, Br)	Br-82	beta, gamma	Kr(36)-82	1.4 days
36 (Krypton, Kr)	Kr-85	beta, gamma	Rb (37)-85	10 years
37 (Rubidium, Rb)	Rb-86	beta, gamma	Sr(38)-86	19 days
37 (Rubidium, Rb)	Rb-87	beta	Sr(38)-87	4.9×10^{10} years
38 (Strontium, Sr)	Sr-89	beta	Y (39)-89	50 days
38 (Strontium, Sr)	Sr-90	beta	Y (39)-90	29 years
39 (Yttrium, Y)	Y-90	beta	Zr(40)-90	2.6 days
40 (Zirconium, Zr)	Zr-95	beta, gamma	Nb(41)-95	64 days
40 (Zirconium, Zr)	Zr-97	beta, gamma	Nb(41)-97	17 hours
41 (Niobium, Nb)	Nb-94	beta, gamma	Mo(42)-94	24,000 years
41 (Niobium, Nb)	Nb-95	beta, gamma	Mo(42)-95	35 days
42 (Molybdenum)	Mo-99	beta, gamma	Tc(43)-99	3 days
43 (Technetium, Tc)	Tc-98	beta, gamma	Ru(44)-98	4,200,000 years
43 (Technetium, Tc)	Tc-99	beta	Ru(44)-99	210,000 years
44 (Ruthenium, Ru)	Ru-103	beta, gamma	Rh (45)-103	39.8 days
44 (Ruthenium, Ru)	Ru-106	beta	Rh (45)-106	1.0 year
45 (Rhodium, Rh)	Rh-106	beta, gamma	Pd(46)-106	35 hours
46 (Palladium, Pd)	Pd-109	beta, gamma	Ag(47)-109	13 hours
47 (Silver, Ag)	Ag-110	beta, gamma	Cd(48)-110	8 months
47 (Silver, Ag)	Ag-111	beta, gamma	Cd(48)-111	7.5 days
48 (Cadmium, Cd)	Cd-115	beta, gamma	In(49)-115	44 days
49 (Indium, In)	In-115	beta	Sn(50)-115	6×10^{14} years
50 (Tin, Sn)	Sn-121	beta	Sb(51)-121	27 hours
51 (Antimony, Sb)	Sb-122	beta, gamma	Te(52)-122	2.7 days
51 (Antimony, Sb)	Sb-124	beta, gamma	Te(52)-124	60 days
51 (Antimony, Sb)	Sb-125	beta, gamma	Te(52)-125	3 years

52 (Tellurium, Te)	Te-127	beta, gamma	I (53)-127	9 hours
52 (Tellurium, Te)	Te-129	beta, gamma	I (53)-129	34 days
53 (Iodine, I)	I-129	beta, gamma	Xe(54)-129	17,000,000 years
53 (Iodine, I)	I-131	beta, gamma	Xe(54)-131	8 days
54 (Xenon, Xe)	Xe-133	beta, gamma	Cs(55)-133	5.3 days
55 (Cesium, Cs)	Cs-134	beta, gamma	Ba(56)-134	2 years
55 (Cesium, Cs)	Cs-135	beta	Ba(56)-135	3,000,000 years
55 (Cesium, Cs)	Cs-137	beta, gamma	Ba(56)-137	30 years
56 (Barium, Ba)	Ba-140	beta, gamma	La(57)-140	12.8 days
57 (Lanthanum, La)	La-140	beta, gamma	Ce(58)-140	40 hours
58 (Cerium, Ce)	Ce-141	beta, gamma	Pr(59)-141	32.5 days
58 (Cerium, Ce)	Ce-143	beta, gamma	Pr(59)-143	1.4 days
58 (Cerium, Ce)	Ce-144	beta, gamma	Pr(59)-144	11.8 days
59(Praseodymium)	Pr-142	beta, gamma	Nd(60)-142	19 hours
59(Praseodymium)	Pr-143	beta, gamma	Nd(60)-143	13.8 days
59(Praseodymium)	Pr-144	beta	Nd(60)-144	17 minutes
60 (Neodymium)	Nd-147	beta, gamma	Pm(61)-147	11 days
61 (Promethium)	Pm-147	beta	Sm(62)-147	2.3 years
61 (Promethium)	Pm-149	beta, gamma	Sm(62)-149	53 hours
61 (Promethium)	Pm-151	beta, gamma	Sm(62)-151	28 hours
62 (Samarium)	Sm-146	alpha	Nd(60)-142	103,000,000 yrs
62 (Samarium)	Sm-153	beta, gamma	Eu(63)-154	1.9 days
63 (Europium, Eu)	Eu-152	beta, gamma	Gd(64)-152	13.4 years
65 (Terbium, Tb)	Tb-160	beta, gamma	Dy(66)-160	2 1/2 months
66 (Dysprosium, Dy)	Dy-154	alpha	Gd(64)-150	3,000,000 years
67 (Holmium, Ho)	Ho-166	beta	Er(68)-166	1.1 days
68 (Erbium, Er)	Er-169	beta, gamma	Tm(69)-169	9.4 days
68 (Erbium, Er)	Er-171	beta, gamma	Tm(69)-171	7.5 hours
69 (Thulium, Tm)	Tm-170	beta	Yb(70)-170	no data
70 (Ytterbium, Yb)	Yb-175	beta, gamma	Lu(71)-175	4.2 days
71 (Lutetium, Lu)	Lu-177	beta, gamma	Hf(72)-177	6.7 days
72 (Hafnium, Hf)	Hf-181	beta, gamma	Ta(73)-181	42 days
72 (Hafnium, Hf)	Hf-182	beta	Ta(73)-182	9,000,000 years
73 (Tantalum, Ta)	Ta-182	beta, gamma	W(74)-182	3.8 months
74 (Tungsten, W)	W-185	beta, gamma	Re(75)-185	75 days
74 (Tungsten, W)	W-187	beta, gamma	Re(75)-187	1 day
75 (Rhenium, Re)	Re-186	beta, gamma	Os(76)-186	3.7 days
75 (Rhenium, Re)	Re-187	beta, gamma	Os(76)-187	4.5×10^{10} years
75 (Rhenium, Re)	Re-188	beta, gamma	Os(76)-188	17 hours
76 (Osmium, Os)	Os-191	beta, gamma	Ir(77)-191	15 days
77 (Iridium, Ir)	Ir-192	beta, gamma	Pt(78)-192	2 1/2 months

78 (Platinum, Pt)	Pt-197	beta, gamma	Au(79)-197	18 hours
79 (Gold, Au)	Au-198	beta, gamma	Hg(80)-198	2.7 days
79 (Gold, Au)	Au-199	beta, gamma	Hg(80)-199	3.1 days
81 (Thallium, Tl)	Tl-204	beta, gamma	Pb(82)-204	3.8 years
81 (Thallium, Tl)	Tl-208	beta, gamma	Pb(82)-208	3 minutes
82 (Lead, Pb)	Pb-210	beta, gamma	Bi(83)-210	22 years
82 (Lead, Pb)	Pb-214	beta, gamma	Bi(83)-214	27 minutes
83 (Bismuth, Bi)	Bi-210	beta	Po(84)-210	5 days
83 (Bismuth, Bi)	Bi-214	beta, gamma	Po(84)-214	20 minutes
84 (Polonium, Po)	Po-209	alpha, gamma	Pb(82)-205	105 years
84 (Polonium, Po)	Po-210	alpha, gamma	Pb(82)-206	138 days
84 (Polonium, Po)	Po-211	alpha, gamma	Pb(82)-207	.5 seconds
84 (Polonium, Po)	Po-214	alpha	Pb(82)-210	.00016 seconds
84 (Polonium, Po)	Po-216	alpha	Pb(82)-212	.15 seconds
84 (Polonium, Po)	Po-218	alpha	Pb(82)-214	3 minutes
85 (Astatine, At)	At-211	alpha	Bi(83)-207	7.2 hours
86 (Radon, Rn)	Rn-219	alpha, gamma	Po(84)-215	4 seconds
86 (Radon, Rn)	Rn-220	alpha, gamma	Po(84)-216	56 seconds
86 (Radon, Rn)	Rn-222	alpha	Po(84)-218	3.8 days
87 (Francium, Fr)	Fr-223	beta, gamma	At(85)-219	22 minutes
88 (Radium, Ra)	Ra-223	alpha, gamma	Rn(86)-219	11 days
88 (Radium, Ra)	Ra-224	alpha, gamma	Rn(86)-220	3.6 days
88 (Radium, Ra)	Ra-226	alpha, gamma	Rn(86)-222	1,600 years
88 (Radium, Ra)	Ra-228	beta	Ac(89)-228	5.75 years
89 (Actinium, Ac)	Ac-225	alpha	Fr(87)-221	10 days
89 (Actinium, Ac)	Ac-227	beta, gamma	Th(90)-227	22 days
89 (Actinium, Ac)	Ac-228	beta, gamma	Th(90)-228	6.1 hours
90 (Thorium, Th)	Th-228	alpha, gamma	Ra(88)-224	1.9 years
90 (Thorium, Th)	Th-229	alpha, gamma	Ra(88)-225	7,300 years
90 (Thorium, Th)	Th-230	alpha, gamma	Ra(88)-226	75,400 years
90 (Thorium, Th)	Th-231	beta, gamma	Pa(91)-231	25.5 hours
90 (Thorium, Th)	Th-232	alpha, gamma	Ra(88)-228	1×10^{10} years
90 (Thorium, Th)	Th-234	beta, gamma	Pa(91)-234	24 days
91 (Protactinium)	Pa-231	alpha, gamma	Ac(89)-227	32,700 years
91 (Protactinium)	Pa-232	beta, gamma	U(92)-232	1.3 days
91 (Protactinium)	Pa-233	beta, gamma	U(92)-233	27 days
91 (Protactinium)	Pa-234	beta, gamma	U(92)-234	7 hours
92 (Uranium, U)	U-234	alpha	Th(90)-230	270,000 yrs
92 (Uranium, U)	U-235	alpha, gamma	Th(90)-231	710,000,000 yrs
92 (Uranium, U)	U-236	alpha, gamma	Th(90)-232	23,400,000 yrs
92 (Uranium, U)	U-238	alpha	Th(90)-234	4,500,000,000 yrs

93 (Neptunium, Np)	Np–237	alpha, gamma	Pa(91)–233	2,140,000 yrs
94 (Plutonium, Pu)	Pu–239	alpha, gamma	U(92)–235	24,110 yrs
94 (Plutonium, Pu)	Pu–242	alpha, gamma	U(92)–238	376,000 yrs
94 (Plutonium, Pu)	Pu–244	alpha, gamma	U(92)–240	82,000,000 yrs

A.5.2 Half-lives listed in order of time

A) Less than a Day

Isotope	Decay Type	Half-Life
Polonium-214	alpha	.00016 seconds
Polonium-216	alpha	.15 seconds
Polonium-211	alpha, gamma	.5 seconds
Radon-219	alpha, gamma	4 seconds
Rhodium-106	beta, gamma	30 seconds
Radon-220	alpha, gamma	56 seconds
Oxygen-15	no data	2.0 minutes
Thallium-208	beta, gamma	3.0 minutes
Polonium-218	alpha	3.1 minutes
Nitrogen-13	beta	10 minutes
Praseodynium-144	beta	17 minutes
Bismuth-214	beta, gamma	20 minutes
Francium-223	beta, gamma	21.8 minutes
Lead-214	beta, gamma	26.8 minutes
Fluorine-18	beta	1.8 hours
Manganese-56	beta, gamma	2.5 hours
Scandium-44	beta, gamma	3.9 hours
Calcium-47	beta, gamma	4.5 hours
Actinium-228	beta, gamma	6.1 hours
Protactinium-234	beta, gamma	6.7 hours
Astatine-211	alpha	7.2 hours
Erbium-171	beta, gamma	7.5 hours
Tellurium-127	beta, gamma	9.5 hours
Germanium-77	beta, gamma	11.3 hours
Potassium-42	beta, gamma	12.4 hours
Copper-64	beta, gamma	12.7 hours
Palladium-109	beta, gamma	13.5 hours
Gallium-72	beta, gamma	13.9 hours
Zirconium-97	beta, gamma	16.8 hours
Rhenium-188	beta, gamma	16.9 hours
Platinum-197	beta, gamma	18.3 hours
Praseodynium-142	beta, gamma	19.1 hours
Magnesium-28	beta, gamma	21.0 hours
Potassium-43	beta, gamma	22.3 hours
Tungsten-187	beta, gamma	23.9 hours (1 day)

B) <u>Within a week</u> (1 day to 7 days)

Isotope	Decay Type	Half-Life
Thorium-231	beta, gamma	1.1 days (25.5 hours)
Arsenic-76	beta, gamma	1.1 days (26.3 hours)
Holmium-166	beta	1.1 days (26.8 hours)
Tin-121	beta	1.1 days (27.0 hours)
Promethium-151	beta, gamma	1.2 days (28.4 hours)
Protactinium-232	beta, gamma	1.3 days (31.3 hours)
Cerium-143	beta, gamma	1.4 days (33.0 hours)
Bromine-82	beta, gamma	1.5 days (35.3 hours)
Rhodium-106	beta, gamma	1.5 days (35.4 hours)
Lanthanum-140	beta, gamma	1.6 days (40.3 hours)
Samarium-153	beta, gamma	1.9 days (46.7 hours)
Promethium-149	beta, gamma	2.2 days (53.1 hours)
Copper-67	beta, gamma	2.6 days (61.9 hours)
Yttrium-90	beta	2.6 days (64.0 hours)
Gold-198	beta, gamma	2.7 days (64.6 hours)
Antimony-122	beta, gamma	2.7 days (65.0 hours)
Molybdenum-99	beta, gamma	2.8 days (65.9 hours)
Gold-199	beta, gamma	3.1 days (75.4 hours)
Scandium-47	beta, gamma	3.4 days (81.6 hours)
Radium-224	alpha, gamma	3.6 days (87.8 hours)
Rhenium-186	beta, gamma	3.7 days (88.9 hours)
Radon-222	alpha	3.8 days (91.7 hours)
Ytterbium-175	beta, gamma	4.2 days (97.9 hours)
Bismuth-210	beta	5.0 days (120.0 hours)
Xenon-133	beta, gamma	5.3 days (127.2 hours)
Lutetium-177	beta, gamma	6.7 days (160.8 hours)

C) <u>1 Week to 1 Month</u> (7 days to 30 days)

Isotope	Decay Type	Half-Life
Ag-111	beta, gamma	7.5 days
Iodine-131	beta, gamma	8.0 days
Sulfur-35	beta	8.7 days
Erbium-169	beta, gamma	9.4 days
Actinium-225	alpha	10.0 days
Neodymium-147	beta, gamma	10.9 days
Radium-223	alpha, gamma	11.4 days
Cerium-144	beta, gamma	11.8 days
Barium-140	beta, gamma	12.8 days
Praseodynium-143	beta, gamma	13.8 days
Phosphorus-32	beta	14.2 days
Osmium-191	beta, gamma	15.4 days
Arsenic-74	beta, gamma	17.8 days
Rubidium-86	beta, gamma	18.6 days
Actinium-227	beta, gamma	21.8 days
Thorium-234	beta, gamma	24.1 days
Phosphorus-33	beta	25.3 days
Protactinium-233	beta, gamma	27.0 days

D) <u>1 Month to 6 months</u> (30 days to 180 days)

Isotope	Decay Type	Half-Life
Cerium-141	beta, gamma	1.1 months (32.5 days)
Tellurium-129	beta, gamma	1.1 months (34.0 days)
Niobium-95	beta, gamma	1.1 months (35.0 days)
Ruthenium-103	beta, gamma	1.2 months (39.2 days)
Hafnium-181	beta, gamma	1.4 months (42.4 days)
Iron-59	beta, gamma	1.4 months (44.5 days)
Cadmium-115	beta, gamma	1.4 months (44.6 days)
Strontium-89	beta	1.6 months (50.5 days)
Antimony-124	beta, gamma	2.0 months (60.4 days)
Zirconium-95	beta, gamma	2.1 months (64.0 days)
Terbium-160	beta, gamma	2.4 months (72.4 days)
Iridium-192	beta, gamma	2.4 months (73.8 days)
Tungsten-185	beta, gamma	2.4 months (74.8 days)
Scandium-46	beta, gamma	2.7 months (83.8 days)
Selenium-75	beta, gamma	3.9 months (118.5 days)
Polonium-210	alpha	4.5 months (138.4 days)
Calcium-45	beta	5.4 months (163.8 days)

E) 6 Months to 1 Year

Isotope	Decay Type	Half-Life
Silver-110	beta, gamma	8 months (250 days)

F) 1 Year to 30 Years:

Isotope	Decay Type	Half-Life
Ruthenium-106	beta	1.02 year (372 days)
Thorium-228	alpha, gamma	1.9 years
Cesium-134	beta, gamma	2.1 years
Promethium-147	beta	2.3 years
Antimony-125	beta, gamma	2.7 years
Thallium-204	beta, gamma	3.8 years
Cobalt-60	beta, gamma	5.3 years
Radium-228	beta	5.7 years
Krypton-85	beta, gamma	10.7 years
Hydrogen-3	beta	12.3 years
Europium-152	beta, gamma	13.4 years
Sodium-24	beta, gamma	14.9 years
Lead-210	beta, gamma	22.3 years
Strontium-90	beta	29.0 years
Cesium-137	beta, gamma	30.2 years

G) 31 yrs - 300 years

Isotope	Decay Type	Half-Life
Silicon-32	beta	100 years
Nickel-63	beta	100 years
Polonium-209	alpha, gamma	105 years
Argon-39	beta	269 years

H) 1,000 years to 10,000 years

Isotope	Decay Type	Half-Life
Radium-226	alpha, gamma	1,600 years
Carbon-14	beta	5,730 years
Thorium-229	alpha, gamma	7,300 years

I) <u>10,000 years to 500,000 years</u>

Isotope	Decay Type	Half-Life
Niobium-94	beta, gamma	24,000 years
Plutonium-239	alpha, gamma	24,110 years
Protactinium-231	alpha, gamma	32,700 years
Thorium-230	alpha, gamma	75,400 years
Iron-60	beta	100,000 years
Technetium-99	beta	210,000 years
Uranium-234	alpha	270,000 years
Chlorine-36	beta	310,000 years
Plutonium-242	alpha, gamma	376,000 years

J) <u>Million years</u>

Isotope	Decay Type	Half-Life
Beryllium-10	beta	1,600,000 years
Neptunium-237	alpha, gamma	2,140,000 years
Cesium-135	beta	3,000,000 years
Dysprosium-154	alpha	3,000,000 years
Technetium-98	beta, gamma	4,200,000 years
Hafnium-182	beta	9,000,000 years
Iodine-129	beta, gamma	17,000,000 years
Uranium-236	alpha, gamma	23,400,000 years
Plutonium-244	alpha, gamma	82,000,000 years
Samarium-146	alpha	103,000,000 years
Uranium-235	alpha, gamma	710,000,000 years

K) <u>Billion years+</u>

Isotope	Decay Type	Half-Life
Potassium-40 (.01%)	beta, gamma	1,280,000,000 years
Uranium-238 (99.3%)	alpha	4,500,000,000 years

(Number in parentheses is % natural abundance)

A.5.3   Half-lives and number of steps

% of sample before decay	Half-life step	% of sample left
100%	1	50%
50%	2	25%
25%	3	12.50%
12.50%	4	6.25%
6.25%	5	3.13%
3.13%	6	1.56%
1.56%	7	0.78%

A.5.4   Multiple Decay Radioactive Isotopes

Most of the radioactive isotopes go through just one decay, turning into one new element and becoming stable. However, there are a some radioactive isotopes that must go through several steps of radioactive isotopes, and steps of radioactive decay, before finally becoming stable. These radioactive isotopes are summarized below.

Original Isotope	# steps	Total Decay
Silicon–32 (Si–32)	2	2 beta
Calcium–47 (Ca–47)	2	2 beta; 2 gamma
Iron–60 (Fe–60)	2	2 beta; 1 gamma
Strontium–90 (Sr–90)	2	2 beta
Zirconium–95 (Zr–95)	2	2 beta; 2 gamma
Molybdenum–99 (Mo–90)	2	2 beta; 1 gamma
Ruthenium–106 (Ru–106)	2	2 beta; 1 gamma
Cadmium–115 (Cd–115)	2	2 beta; 1 gamma
Tellurium–129 (Te–129)	2	2 beta; 2 gamma
Barium–140 (Ba–140)	2	2 beta; 2 gamma
Cerium–144 (Ce–143)	2	2 beta; 1 gamma
Neodymium–147(Nd–147)	2	2 beta; 1 gamma
Hafnium–182 (Hf–182)	2	2 beta; 1 gamma
Tungsten–187 (W–187)	2	2 beta; 2 gamma
Lead–210 (Pb–210)*	3	2 beta; 2 gamma; 1 alpha
Lead–214 (Pb–214)*	6	2 alpha; 4 beta; 4 gamma
Bismuth–210 (Bi–210)*	2	1 alpha; 1 beta; 1 gamma
Bismuth–214 (Bi–214)*	5	3 beta; 3 gamma; 2 alpha
Polonium–214 (Po–214)*	4	2 alpha; 2 beta; 2 gamma
Polonium–218 (Po–218)*	7	3 alpha; 4 beta; 4 gamma
Radon–220 (Rn–220)**	2	2 alpha; 1 gamma
Radon–222 (Rn–222)*	8	4 alpha; 4 beta; 4 gamma
Radium–223 (Ra–223)	2	2 alpha; 2 gamma
Radium–224 (Ra–224)**	3	3 alpha; 2 gamma
Radium–226 (Ra–226)*	9	5 alpha; 4 beta; 5 gamma
Radium–228 (Ra–228)**	6	4 alpha; 2 beta; 4 gamma
Actinium–228 (Ac–228)**	5	4 alpha; 1 beta; 4 gamma
Thorium–228 (Th–228)**	4	4 alpha; 3 gamma
Thorium–230 (Th–230)*	10	6 alpha; 4 beta; 6 gamma
Thorium–231 (Th–231)***	3	1 alpha; 2 beta; 3 gamma
Thorium–232 (Th–232)**	7	5 alpha; 2 beta; 5 gamma
Thorium–234 (Th–234)****	13	7 alpha; 6 beta; 8 gamma

Protactin.–231 (Pa–231)***	2	1 alpha; 1 beta; 2 gamma
Protactin.–234 (Pa–234)****	12	7 alpha; 5 beta; 7 gamma
Uranium–234 (U–234)*	11	7 alpha; 4 beta; 6 gamma
Uranium–235 (U–235)***	2	2 alpha; 2 beta; 4 gamma
Uranium–236 (U–236)**	8	6 alpha; 2 beta; 6 gamma
Uranium–238 (U–238)****	14	8 alpha; 6 beta; 8 gamma
Neptunium–237 (Np–237)	2	1 alpha; 1 beta; 2 gamma
Plutonium–238 (Pu–238)*	12	8 alpha; 4 beta; 7 gamma
Plutonium–239 (Pu–239)***	5	3 alpha; 2 beta; 5 gamma
Plutonium–240 (Pu–240)**	9	7 alpha; 2 beta; 7 gamma
Plutonium–242 (Pu–242)****	15	9 alpha; 6 beta; 9 gamma

*For detailed steps see decay sequence of Pu–238 or Pu–242 (A 5.5)
**For detailed steps see decay sequence of Pu–240 (Appendix 5.5)
***For detailed steps see decay sequence of Pu–239 (Appendix 5.5)
****For detailed steps see decay sequence of Pu–242 (Appendix 5.5)

A.5.5   Decay Sequences for Multiple Decay Isotopes

Listed below are the detailed sequences for four multi-step radioisotopes (Pu-238, Pu-239, Pu-240, and Pu-242.) Note that these sequences actually have other (shorter) multi-step decay sequences within them. Listing the sequences in this way we can save space in the book and yet provide all the data that we need. For example, to find the decay sequence for Th-230, look within Plutonium-238 sequence below, starting at step 3. The sequence for Th-230 decay continues from there.

Plutonium-238 (Pu-238) Decay Sequence

1. Pu-238 becomes U-234; alpha, gamma; 87.74 years
2. U-234 becomes Th-230; alpha; 270,000 years
3. Th-230 becomes Ra-226; alpha, gamma; 75,400 years
4. Ra-226 becomes Rn-222; alpha, gamma; 1,600 years
5. Rn-222 becomes Po-218; alpha; 3.8 days
6. Po-218 becomes Pb-214; alpha; 3 minutes
7. Pb-214 becomes Bi-214; beta, gamma; 27 minutes
8. Bi-214 becomes Po-214; beta, gamma; 20 minutes
9. Po-214 becomes Pb-210; alpha; .00016 seconds
10. Pb-210 becomes Bi-210; beta, gamma; 22 years
11. Bi-210 becomes Po-210; beta; 5 days
12. Po-210 becomes Pb-206; alpha, gamma; 138 days

Plutonium-239 (Pu-239) Decay Sequence

1. Pu-239 becomes U-235; alpha, gamma; 24,110 years
2. U-235 becomes Th-231; alpha, gamma, 710,000,000 years
3. Th-231 becomes Pa-231; beta, gamma; 25.5 hours
4. Pa-231 becomes Ac-227; alpha, gamma; 32,700 years
5. Ac-227 becomes Th-227; beta, gamma; 22 days

Plutonium-240 (Pu-240) Decay Sequence

1. Pu-240 becomes U-236; alpha, gamma; 6,537 years
2. U-236 becomes Th-232; alpha, gamma; 23,400,000 years
3. Th-232 becomes Ra-228; alpha, gamma; 1×10^{10} years
4. Ra-228 becomes Ac-228; beta; 5.75 years
5. Ac-228 becomes Th-228; beta, gamma; 6.1 hours
6. Th-228 becomes Ra-224; alpha, gamma; 1.9 years
7. Ra-224 becomes Rn-220; alpha, gamma, 3.6 days
8. Rn-220 becomes Po-216; alpha, gamma; 56 seconds
9. Po-216 becomes Pb-212; alpha; .15 seconds

Plutonium-242 (Pu-242) Decay Sequence

1. Pu-242 becomes U-238; alpha, gamma; 376,000 years
2. U-238 becomes Th-234; alpha; 4,500,000,000 years
3. Th-234 becomes Pa-234; beta, gamma; 24 days
4. Pa-234 becomes U-234; beta, gamma; 7 hours
5. U-234 becomes Th-230; alpha; 270,000 years*
6. Th-230 becomes Ra-226; alpha, gamma; 75,400 years
7. Ra-226 becomes Rn-222; alpha, gamma; 1,600 years
8. Rn-222 becomes Po-218; alpha; 3.8 days
9. Po-218 becomes Pb-214; alpha; 3 minutes
10. Pb-214 becomes Bi-214; beta, gamma; 27 minutes
11. Bi-214 becomes Po-214; beta, gamma; 20 minutes
12. Po-214 becomes Pb-210; alpha; .00016 seconds
13. Pb-210 becomes Bi-210; beta, gamma; 22 years
14. Bi-210 becomes Po-210; beta; 5 days
15. Po-210 becomes Pb-206; alpha, gamma; 138 days

*Note that the decay steps from this point forward are also found in the sequence for Pu-238.

A.5.6   Radioactive Decay Summaries

% of Radioisotopes that do Each Decay
· 20% of radioactive isotopes do alpha decay
· 80% of radioactive isotopes do beta decay
· 70% of radioactive isotopes do gamma decay*
*Gamma must accompany either alpha or beta; gamma does not occur by itself.

% of Radioisotopes with half-life times less than 30 years
a. Less than 30 years: 80%
b. Greater than 30 years: 20%

Total time to reach less than 1%	% of radioisotopes requiring that time
Less than 1 week	25%
1 week – 1 month	17%
1 week – 1 year	20%
1 year – 5 years	7%
5 years – 50 years	6%*
50 years – 250 years	5%
250 years – 1,000 years	2%
1,000 years – 1 million years	6%
1 million years – 1 billion years	10%
1 billion years+	2%**

*Subtotal #1: 75% of all radioactive isotopes will decay to less than 1% of the original amount within 50 years.

**Subtotal #2: 25% of all radioactive isotopes will require 50 years or longer to decay to less than 1% of the original amount.

A.5.7   American Nuclear Society Standards

The following ANS safety standards are some I recommend for citizens to read. They can be found at www.ans.org/store/vc-stnd.

Plant Design and Construction

#240223: Design of Concrete Radiation Shielding for Nuclear Power Plants

#240247: Containment System Leakage Testing Requirements

#240177: Auxiliary Feedwater System for Pressurized Water Reactors

#240207: Design Criteria for Safe Shutdown Following Selected Design Basis Events in Light Water Reactors

Training and Operation

#240219: Administrative Practices for Nuclear Criticality Safety

#240188: Qualification & Training of Personnel for Nuclear Power Plants

#240231: Simulators for Use in Operator Training and Examination

#240230: Criteria for Planning, Development, Conduct, and Evaluation of Drills and Exercises for Emergency Preparedness

Radioactive Waste Storage

#240193: Solid Radioactive Waste Processing for Water Cooled Reactor Plants

#240170: Design Criteria for Spent Fuel Storage Installation (Water Pool Type)

#240185: Design Criteria for Spent Fuel Storage Installation (Dry Type)

Earthquakes, Tornadoes

#240105: Assessing Capability for Surface Faulting at Power Reactor Sites

#240122: Estimating Extreme Wind Characteristics at Nuclear Power Sites

#240007: Evaluating Site-Related Geotechnical Parameters at Nuclear Power Sites

#240129: Determining Meteorological Information at Nuclear Power Sites

#240008: Combining Natural and External Man-Made Hazards at Reactor Sites

#240006: Processing Records Obtained From Seismic Instrumentation

#240251: Handling Records from Nuclear Power Plant Seismic Instrumentation

Underground Water Supplies

#240005: Evaluation of Ground Water Supply for Nuclear Power Sites

#240009: Evaluation of Surface-Water Supplies for Nuclear Power Sites

A.5.8   NRC Regulation Guides

The following NRC regulation guides are some I recommend for citizens to read. They can be found at www.nrc.gov/reading-rm/doc-collections.

Fact Sheets
- Fact Sheet on Nuclear Power Plant Licensing Process
- Fact Sheet on Reactor Operator Licensing

Plant Design: Site locations, Foundations, Structure
- Regulatory Guide 1.132: Site Investigations for Foundations of Nuclear Power Plants
- Regulatory Guide 4.7: General Site Suitability Criteria for Nuclear Power Stations
- Regulatory Guide 1.142: Safety-Related Concrete Structures for Nuclear Power Plants (Other Than Reactor Vessels and Containments)

Plant Design: Water and Piping
- Regulatory Guide 1.178: An Approach for Plant-Specific Risk-Informed Decision Making for In-service Inspection of Piping
- Regulatory Guide 1.182: Water Sources for Long-Term Recirculation Cooling Following a Loss-of-Coolant Accident

Training and Operation
- Regulatory Guide 1.8: Training of Personnel for Nuclear Power Plants
- Regulatory Guide 1.149: Nuclear Power Plant Simulation Facilities for Use in Operator Training and License Examinations
- Regulatory Guide 1.101: Emergency Planning for Nuclear Power Reactors

Bibliography

Nuclear Power

1. Energy for Man: From Windmills to Nuclear Power, by Hans Thirring, 1958. Indiana University Press.
2. Energy Resources, by Andrew Simon, 1975. Publisher: Pergamon Press, Inc
3. Nontechnical Guide to Energy Resources, by Ben Ebenhack, 1995. PennWell Publishing Company
4. Electric Power Generation: A Nontechnical Guide, by Barnett and Bjornsgaard, 2000. Publisher: PennWell Publishing Company
5. Energy: A Guidebook, by Janet Ramage, 1997. Oxford University Press.
6. The Elements, Second Edition, by John Emsley, 1990. Oxford University Press
7. Chemistry, by Wilbraham, Staley, Simpson, and Matta, 1987. Addison-Wesley
8. Web Elements www.webelements.com
9. Jefferson Lab's Elements http://education.jlab.org/itselemental/index.html
10. The Secrets of Uranium: An Internet Hotlist of Uranium, updated Feb 2005. http://www.kn.sbc.com/wired/fil/pages/listuraniumco.html
11. The Institute for Energy and Environmental Research (IEER) www.ieer.org/index.html
12. IEER Fact Sheets www.ieer.org/fctsheet/index.html
13. Table of Isotopes, Berkeley National Laboratory, http://ie.lbl.gov/education/isotopes.htm
14. Chemistry in Context, Second Edition, by the American Chemical Society (various contributors), 1997. Publisher: American Chemical Society
15. Totse, Informational website http://www.totse.com
16. Powerhouse: Inside a Nuclear Power Plant, by Charlotte Wilcox and Jerry Boucher, 1997. Publisher: Carolrhoda Books, Inc.
17. Nuclear Regulatory Commission (NRC) http://www.nrc.gov
18. Effects of Nuclear Radiation http://slhscif.tripod.com/benchmark1/documents/Yihan_Zhao.htm
19. Dosages of Nuclear Radiation to Humans http://www.psigate.ac.uk/newsite/reference/plambeck/chem1/p05015.htm
20. Effects of Nuclear Radiation on the Human Body http://www.millennium-ark.net/News_Files/NBC/radiation.human.body.html
21. Toxicology: The Basic Science of Poisons, 4th edition, edited by Cassarett, Doull, Ambur, and Klassen, 1993. Publisher: McGraw-Hill.
22. Nuclear Accidents, by Joel Helgerson, 1988. Publisher: Impact Books.
23. Three Mile Island, by Michael Cole, 2002. Publisher: Enslow Publishers, Inc.
24. Three Mile Island: A Nuclear Crisis in Historical Perspective, by J. Samuel Walker, 2004. University of California Press.

25. The Chernobyl Catastrophe, by Graham Rickard, 1989. Bookwright Press

26. Emergency Responder Training Manual for the Hazardous Materials Technician, by The Center for Labor Education and Research (CLER), edited by Lori Andrews, 1992. Publisher: Van Nostrand Reinhold.

27. The Demise of Nuclear Energy? Lessons for Democratic Control of Technology, by Morone and Woodhouse, 1989, Yale University Press

28. International Nuclear Safety Program http://insp.pnl.gov

29. International Nuclear Safety Center, Dpt of Energy www.insc.anl.gov

30. Atomic Insights www.atomicinsights.com

31. Yucca Mountain Project, Office of Civilian Radioactive Waste Management, Dpt of Energy http://www.ocrwm.doe.gov/ymp/about/index.shtml

32. Yucca Mountain Analyzed http://greennature.com/article273.html

33. "A New Vision for Nuclear Waste" by Matthew Wald, *MIT's Technology Review*, Dec. 2004.

34. "Heavy-Metal Nuclear Power" by Dr. Eric Loewen, *American Scientist*, Nov-Dec 2004.

35. Introduction to the Principles of Ceramic Processing, by James Reed, 1988. John Wiley & Sons.

36. Introduction to Phase Equilibria in Ceramics, by Bergeron and Risbud, 1984. The American Ceramic Society.

37. "Waste Mismanagement", by Christopher Helman, *Forbes Magazine*, August 15, 2005

38. American Nuclear Society (ANS) www.ans.org

39. Nuclear Safety Standards, American Nuclear Society www.ans.org/store

40. NRC Regulation Guides www.nrc.gov/reading-rm/doc-collections

41. Industrial Fire Hazards Handbook, National Fire Protection Association, 1979.

Toxicology and Safety

1. Toxicology: The Basic Science of Poisons, 4th edition, edited by Cassarett, Doull, Ambur, and Klassen, 1993. McGraw-Hill.

2. Emergency Responder Training Manual for the Hazardous Materials Technician, by The Center for Labor Education and Research (CLER), edited by Lori Andrews, 1992. Publisher: Van Nostrand Reinhold.

3. Industrial Fire Hazards Handbook, National Fire Protection Association, 1979.

4. Toxics Release Inventory Program (TRI) www.epa.gov/tri

5. The Extension Toxicology Network (EXTOXNET) http://extoxnet.orst.edu

6. NIOSH Databases www.cdc.gov/niosh/database.html

7. NIOSH Pocket Guide to Chemical Hazards (NPG) www.cdc.gov/niosh/npg/npg.html

8. IPCS (International Programme on Chemical Safety), via NIOSH www.cdc.gov/niosh/ipcs/icstart.html

9. NIOSH Occupational Health Guidelines for Chemical Hazards
   http://www.cdc.gov/niosh/81-123.html
10. OSHA (Occupational Safety and Health Administration) www.osha.gov
11. Eureka County Yucca Mountain Information Office www.yuccamountain.org/

Government Sites – General

1. US Department of Energy (DOE) www.energy.gov
2. US Department of the Interior www.doi.gov
3. US Bureau of Reclamation www.usbr.gov
4. US Department of Agriculture (USDA) www.usda.gov
5. Environmental Protection Agency (EPA) www.epa.gov
6. Food and Drug Administration (FDA) www.cfsan.fda.gov
7. National Institute for Occupational Safety and Health (NIOSH)
   www.cdc.gov/niosh
8. Mine Safety and Health Administration (MSHA) www.msha.gov
9. Federal Energy Regulatory Commission (FERC) www.ferc.gov
10. Nuclear Regulatory Commission (NRC) www.nrc.gov
11. National Climatic Data Center (NCDC) www.ncdc.noaa.gov

Department of Energy (DOE) Related Sites

1. Department of Energy (DOE) www.energy.gov
2. Energy Information Administration (EIA) www.eia.doe.gov
3. [Office of] Efficiency and Renewable Energy (EERE) www.eere.energy.gov
4. Office of Fossil Energy (in Dept of Energy) www.fossil.energy.gov
5. Electric Transmission and Distribution Office www.electricity.doe.gov
6. Science (Office of Science) www.sc.doe.gov
7. Nuclear Regulatory Commission (NRC) www.nrc.gov
8. Civilian Radioactive Waste Management (OCRWM) www.ocrwm.doe.gov
9. Yucca Mountain Project www.ocrwm.doe.gov/ymp/about/index.shtml
10. International Nuclear Safety Program http://insp.pnl.gov
11. International Nuclear Safety Center, Argonne Laboratory www.insc.anl.gov
12. National Energy Technology Laboratory (NETL) www.netl.doe.gov
13. National Renewable Energy Laboratory (NREL) www.nrel.gov
14. Oak Ridge National Laboratory www.ornl.gov
15. Los Alamos National Laboratory (LANL) www.lanl.gov/worldview
16. Pacific Northwest National Laboratory (PNL) www.pnl.gov
17. Starlight, from PNNL/DOE http://starlight.pnl.gov

Index

Made in the USA
Coppell, TX
24 May 2024

32740112R10085